Fire Investigator

Principles and Practice to NFPA 921 and 1033
THIRD EDITION

Student Workbook

Jones & Bartlett Learning
World Headquarters
40 Tall Pine Drive
Sudbury, MA 01776
978-443-5000
info@jblearning.com
www.jblearning.com

**International Association
of Fire Chiefs**
4025 Fair Ridge Drive
Fairfax, VA 22033
www.IAFC.org

**International Association
of Arson Investigators**
2111 Baldwin Avenue, Suite 203
Crofton, MD 21114
firearson.com

**National Fire Protection
Association**
1 Batterymarch Park
Quincy, MA 02169-7471
www.NFPA.org

Jones & Bartlett Learning Canada
6339 Ormindale Way
Mississauga, Ontario L5V 1J2
Canada

Jones & Bartlett Learning
International
Barb House, Barb Mews
London W6 7PA
United Kingdom

Jones & Bartlett Learning books and products are available through most bookstores and online booksellers. To contact Jones & Bartlett Learning directly, call 800-832-0034, fax 978-443-8000, or visit our website, www.jblearning.com.

> Substantial discounts on bulk quantities of Jones & Bartlett Learning publications are available to corporations, professional associations, and other qualified organizations. For details and specific discount information, contact the special sales department at Jones & Bartlett Learning via the above contact information or send an email to specialsales@jblearning.com.

Production Credits
Chairman, Board of Directors: Clayton Jones
Chief Executive Officer: Ty Field
President: James Homer
Sr. V.P., Chief Operating Officer: Don W. Jones, Jr.
V.P., Design and Production: Anne Spencer
V.P., Manufacturing and Inventory Control: Therese Connell
V.P., Sales, Public Safety Group: Matthew Maniscalco
Executive Publisher: Kimberly Brophy
Executive Acquisitions Editor—Fire: William Larkin
Associate Production Editor: Lisa Cerrone
Marketing Manager: Brian Rooney
Composition: Glyph International
Cover Design: Kristin E. Parker
Cover Image: Courtesy of Captain David Jackson, Saginaw Township Fire Department
Printing and Binding: Courier Kendallville
Cover Printing: Courier Kendallville

Editorial Credits
Author: Douglas C. Ott

Copyright © 2012 by Jones & Bartlett Learning, LLC and the National Fire Protection Association®

All rights reserved. No part of the material protected by this copyright may be reproduced or utilized in any form, electronic or mechanical, including photocopying, recording, or by any information storage and retrieval system, without written permission from the copyright owner.

The procedures and protocols in this book are based on the most current recommendations of responsible sources. The National Fire Protection Association (NFPA), International Association of Fire Chiefs (IAFC), International Association of Arson Investigators (IAAI), and the publisher, however, make no guarantee as to, and assume no responsibility for, the correctness, sufficiency, or completeness of such information or recommendations. Other or additional safety measures may be required under particular circumstances.

Notice: The individuals described in "Fire Alarms" throughout this text are fictitious.

Photo Credit: Pages 36 and 182, Part F Courtesy of APA – The Engineered Wood Association. Unless otherwise indicated, all photographs and illustrations are under copyright of Jones & Bartlett Learning, LLC.

ISBN: 978-0-7637-7698-5

6048

Printed in the United States of America
15 14 13 12 11   10 9 8 7 6 5 4 3 2 1

Contents

Chapter 1 Administration   2

Chapter 2 Basic Fire Methodology   8

Chapter 3 Basic Fire Science   14

Chapter 4 Fire Patterns   22

Chapter 5 Building Systems   28

Chapter 6 Electricity and Fire   38

Chapter 7 Fuel Gas Systems   46

Chapter 8 Fire-Related Human Behavior   54

Chapter 9 Legal Considerations   60

Chapter 10 Safety   66

Chapter 11 Sources of Information   72

Chapter 12 Planning and Preplanning the Investigation   78

Chapter 13 Documentation of the Investigation   84

Chapter 14 Physical Evidence   90

Chapter 15 Origin Determination   96

Chapter 16 Fire Cause Determination   102

Chapter 17 Analyzing the Incident for Cause and Responsibility   108

Chapter 18 Failure Analysis and Analytical Tools   114

Chapter 19 Explosions   120

Chapter 20 Incendiary Fires   126

Chapter 21 Fire and Explosion Deaths and Injuries   132

Chapter 22 Appliances   138

Chapter 23 Motor Vehicle Fires   144

Chapter 24 Wildfire Investigations   150

Chapter 25 Management of Complex Investigations   158

Chapter 26 Marine Fire Investigations   164

Answer Key

Chapter 1 Administration   171

Chapter 2 Basic Methodology   172

Chapter 3 Basic Fire Science   174

Chapter 4 Fire Patterns   176

Chapter 5 Building Systems   177

Chapter 6 Electricity and Fire   183

Chapter 7 Fuel Gas Systems   187

Chapter 8 Fire-Related Human Behavior   190

Chapter 9 Legal Considerations   192

Chapter 10 Safety   194

Chapter 11 Sources of Information   196

Chapter 12 Planning and Preplanning the Investigation   201

Chapter 13 Documentation of the Investigation   202

Chapter 14 Physical Evidence   204

Chapter 15 Origin Determination   207

Chapter 16 Fire Cause Determination   210

Chapter 17 Analyzing the Incident for Cause and Responsibility   212

Chapter 18 Failure Analysis and Analytical Tools   213

Chapter 19 Explosions   215

Chapter 20 Incendiary Fires   217

Chapter 21 Fire and Explosion Deaths and Injuries   219

Chapter 22 Appliances   222

Chapter 23 Motor Vehicle Fires   225

Chapter 24 Wildfire Investigations   228

Chapter 25 Management of Complex Investigations   230

Chapter 26 Marine Fire Investigations   232

Student Resources

Exam Prep: Fire Investigator, Second Edition
ISBN: 978-1-4496-0962-7

Exam Prep: Fire Investigator, Second Edition is designed to thoroughly prepare you for a Fire Investigation certification, promotion, or training examination by including the same type of multiple-choice questions that you are likely to encounter on the actual exam.

To help improve examination scores, this prep guide follows the Performance Training Systems, Inc. Systematic Approach to Examination Preparation. *Exam Prep: Fire Investigator, Second Edition* was written by fire personnel explicitly for fire personnel, and all content was verified with the latest reference materials and by a technical review committee.

Your exam performance will improve after using this system!

Fire Investigator Field Guide, Second Edition
ISBN-13: 978-0-7637-5852-3

Fire Investigator Field Guide, Second Edition is your direct link to the information you need to conduct thorough and accurate investigations. As a fire investigator, your job is to provide answers as to origin and cause. *Fire Investigator Field Guide, Second Edition* is like having your own personal assistant on hand to locate the facts and figures for you. Save time and get better results with a compact reference library in a single volume!

This substantive resource has tables, charts, lists, art, and more from the most respected references in the field, including NFPA 921 and NFPA 170, NFPA's *Fire Protection Handbook*, and the Society of Fire Protection Engineers' *Handbook of Fire Protection Engineering*.

Data is organized into sections for fast and easy information retrieval, and complete backup is provided for every phase of the investigation process. From pre-arrival activities to documentation and analysis, this guide has you covered!

Technology Resources

www.Fire.jbpub.com

This site has been specifically designed to complement *Fire Investigator: Principles and Practice to NFPA 921 and 1033, Third Edition* and is regularly updated. Resources available include:

- **Chapter Pretests** that prepare students for training. Each chapter has a pretest and provides instant results, feedback on incorrect answers, and page references for further study.
- **Interactivities** that allow students to reinforce their understanding of the most important concepts in each chapter.
- **Hot Term Explorer,** a virtual dictionary, allows students to review key terms, test their knowledge of key terms through quizzes and flashcards, and complete crossword puzzles.

Administration

Workbook Activities

The following activities have been designed to help you. Your instructor may require you to complete some or all of these activities as a regular part of your investigator training program. You are encouraged to complete any activity that your instructor does not assign as a way to enhance your learning in the classroom.

Chapter Review

The following exercises provide an opportunity to refresh your knowledge of this chapter.

Matching

Match each of the terms in the left column to the appropriate definition in the right column.

_____ 1. NFPA Guide
_____ 2. NFPA Standard
_____ 3. AHJ
_____ 4. JPR
_____ 5. NFPA 1033
_____ 6. NFPA 921
_____ 7. NFPA documents
_____ 8. Technical committees
_____ 9. NFPA staff
_____ 10. Call for proposals

A. Authority having jurisdiction
B. Guide for Fire and Explosion Investigations
C. Responsible for NFPA document content
D. Responsible for ensuring committees follow rules governing committee activities
E. The first step in the revision cycle of an NFPA document
F. Does not include terms such as *shall* or *must*
G. Standard for Professional Qualifications for Fire Investigator
H. Designed to establish minimum JPRs
I. Job performance requirements
J. Developed through a consensus process

Multiple Choice

Read each item carefully and then select the best response.

_____ 1. Establishes guidelines and recommendations for the safe and systematic investigation or analysis of fire and explosion incidents:
   A. NFPA 1041
   B. NFPA 1500
   C. NFPA 921
   D. NFPA 1033

_____ 2. Designed to establish the minimum JPRs for service as a fire investigator:
   A. NFPA 1041
   B. NFPA 1500
   C. NFPA 921
   D. NFPA 1033

_____ 3. A document in which the main text contains only mandatory provisions and is in a form that is generally suitable for adoption into law:
   A. Guide
   B. Standard
   C. Legal Note
   D. Memo

_____ 4. A document that is advisory or informatory and that contains nonmandatory provisions:
   A. Guide
   B. Standard
   C. Legal Note
   D. Memo

_____ 5. NFPA documents do not have the power of law unless adopted by a(n)
   A. EPA
   B. DOD
   C. AHJ
   D. CPA

_____ 6. Fire investigators must keep training current by
   A. Attending seminars
   B. Attending workshops
   C. Trade magazines
   D. All of the above

_____ 7. NFPA documents are developed through
   A. Fire chiefs meetings
   B. Consensus process
   C. Congressional law
   D. NFPA staff

_____ 8. All proposed changes to NFPA documents are reviewed by
   A. NFPA staff
   B. Technical committees
   C. Writers guild
   D. NFA

_____ 9. NFPA documents are updated on a cycle of
   A. 3 to 5 years
   B. 1 to 3 years
   C. 5 to 10 years
   D. 10 to 15 years

_____ 10. Proposed NFPA changes can be submitted by
   A. NFPA members only
   B. Chief of the department
   C. Manufacturers and users
   D. Anyone

Fill-in

Read each item carefully and then complete the statement by filling in the missing word(s).

1. NFPA 921 can be used by anyone who is charged with _____ and _____ fires and explosions.

2. The purpose of NFPA 921 is to _____ _____ and _____ for the safe and systematic investigation or analysis of fire and explosion incidents.

3. _____ _____ is a standard that is designed to establish minimum JPRs for fire investigators.

4. NFPA is the publisher of almost 300 _____, _____, _____ _____, and _____.

5. NFPA _____ committees are responsible for document content.

6. Typically, individual documents are updated on regular cycles of _____ to _____ years.

7. The first step for any NFPA document that is entering its revision cycle is issuing a _____ _____ _____.

Vocabulary

Define the following terms using the space provided.

1. Guide:

2. Standard:

3. Fire Investigation:

4. NFPA 921 Guide for Fire and Explosion Investigations:

5. NFPA 1033 Professional Qualifications for Fire Investigator:

Short Answer

Complete this section with short written answers, using the space provided.

1. Describe the minimum qualifications for a fire investigator.

2. Identify 13 topics of which fire investigators should maintain up-to-date knowledge.

3. When does an NFPA document have the power of law?

4. Describe the differences between a standard and a guide.

5. Identify nine membership categories the NFPA uses to fill and balance a committee.

Fire Alarms

The following case scenario will give you an opportunity to explore the concerns associated with fire investigation. Read the scenario and then answer the question in detail.

You have been a fire investigator for several years and have disagreed with parts of NFPA 921 throughout your career. The document is entering its revision cycle and you wish to submit a proposal for change.

1. How would you proceed?

Basic Fire Methodology

Workbook Activities

The following activities have been designed to help you. Your instructor may require you to complete some or all of these activities as a regular part of your investigator training program. You are encouraged to complete any activity that your instructor does not assign as a way to enhance your learning in the classroom.

Chapter Review

The following exercises provide an opportunity to refresh your knowledge of this chapter.

Matching
Match each of the terms in the left column to the appropriate definition in the right column.

_____ 1. Hypothesis

_____ 2. Empirical data

_____ 3. Deductive reasoning

_____ 4. Scientific method

_____ 5. Cognitive testing

_____ 6. Expectation bias

_____ 7. Inductive reasoning

_____ 8. Technical review

A. The systematic pursuit of knowledge involving the recognition and formulation of a problem, the collection of data through observation and experiment, and the formulation and testing of a hypothesis.

B. The process by which conclusions are drawn by logical inference from given premises. Use of knowledge, skills, and art to challenge or test the hypothesis analytically.

C. Preconceived determination or premature conclusions as to what the cause of the fire was.

D. The use of a person's thinking skills and judgment to evaluate the empirical data and challenge the conclusions of the final hypothesis.

E. Data that are collected based on observation or experience and are capable of being verified.

F. The process by which a person starts from a particular experience and proceeds to generalizations. The process by which hypotheses are developed based on observable or known facts and the training, experience, knowledge, and expertise of the observer.

G. Theory supported by the empirical data that the investigator has collected through observation and then developed into explanations for the event, which are based on the investigator's knowledge, training, experience, and expertise.

H. A critique of the investigator's works and findings.

CHAPTER 2

Multiple Choice

Read each item carefully and then select the best response.

_____ 1. What systematic pursuit of knowledge involves recognizing and identifying a particular problem, defining the problem, collecting information, analyzing the information, and developing and testing a hypothesis?
   A. SWAG method
   B. Scientific method
   C. Systematic method
   D. Fire cause method

_____ 2. With what method do you gather data by working from the least burn area to the most burn area or from the outside in toward the origin?
   A. SWAG method
   B. Scientific method
   C. Systematic method
   D. Fire cause method

_____ 3. What is the first step of the scientific method?
   A. Realize there is a problem to be resolved
   B. Define the problem
   C. Collect data
   D. Develop a hypothesis

_____ 4. What is the second step of the scientific method?
   A. Realize there is a problem to be resolved
   B. Define the problem
   C. Collect data
   D. Develop a hypothesis

_____ 5. With what should the collection of data begin?
   A. Define the problem
   B. Take quality pictures
   C. Recognize the need for data
   D. Systematically evaluate the scene

_____ 6. What type of data is collected using the scientific method to be analyzed later?
   A. Analytical data
   B. Hypothetical data
   C. Empirical data
   D. Realistic data

_____ 7. Developing a hypothesis (or hypotheses) based on the empirical data collected is referred to as what?
   A. Inductive reasoning
   B. Empirical reasoning
   C. Hypothetical reasoning
   D. Realistic reasoning

_____ 8. Data that are known to be true and that can be confirmed are referred to as
   A. Analytical data
   B. Hypothetical data
   C. Empirical data
   D. Realistic data

_____ 9. What is the sixth step of the scientific method?
   A. Realize there is a problem to be resolved
   B. Test the hypothesis
   C. Collect data
   D. Develop a hypothesis

_____ 10. What step of the scientific method not only involves the critical examination and expected challenge of others but also requires critical evaluation by the investigator?
   A. Realize there is a problem to be resolved
   B. Test the hypothesis
   C. Collect data
   D. Develop a hypothesis

Fill-in

Read each item carefully and then complete the statement by filling in the missing word(s).

1. Having one or more knowledgeable peers review your investigative process and conclusions and, when appropriate, pose challenges to conclusions offered helps in _____ _____ _____.

2. The investigator has to approach every incident with an open mind and without presumptions or _____ _____.

3. The _____ _____ requires the reviewer to have access to all data on which the investigator relied.

4. The _____ _____ is usually carried out within the investigator's organization to ensure the product meets the organization's procedures.

5. In situations in which the hypothesis is not supportable to a probability and is only _____ or _____, the results should be reflected as such.

6. Testing the hypothesis not only involves the critical examination and expected challenge of others but also requires critical evaluation by the _____.

7. Through _____ _____, the ultimate conclusion is supported, unsupported, or refuted by the complete body of evidence and data.

8. A proper and unbiased _____ _____ usually provides the best means to assess the validity of the investigator's conclusion.

Chapter 2 : Basic Fire Methodology

Vocabulary
Define the following terms using the space provided.

1. Deductive reasoning:
   _____
   _____
   _____

2. Empirical data:
   _____
   _____
   _____

3. Hypothesis:
   _____
   _____
   _____

4. Scientific method:
   _____
   _____
   _____

Short Answer
Complete this section with short written answers, using the space provided.

1. Describe the three levels of hypothesis certainty.
   _____
   _____
   _____
   _____
   _____

2. List eight types of questions the investigator should use in testing the hypothesis.
   _____
   _____
   _____
   _____
   _____

3. List seven categories of data an investigator should gather to help answer the problem that has been defined.

4. List the steps that should be followed in using the scientific method for fire investigations.

5. Identify three types of reviews an investigator's reports should undergo.

Fire Alarms

The following case scenario will give you an opportunity to explore the concerns associated with fire investigation. Read the scenario and then answer the question in detail.

As a new investigator, you have just been assigned your first structure fire investigation. The structure is a single-family two-story dwelling that has sustained fire damage to the front half of the home. Both floors were damaged.

1. Describe the steps that should be followed in using the scientific method for your fire investigation.

   _____
   _____
   _____
   _____
   _____
   _____
   _____
   _____
   _____
   _____
   _____
   _____

Basic Fire Science

Workbook Activities

The following activities have been designed to help you. Your instructor may require you to complete some or all of these activities as a regular part of your investigator training program. You are encouraged to complete any activity that your instructor does not assign as a way to enhance your learning in the classroom.

Chapter Review

The following exercises provide an opportunity to refresh your knowledge of this chapter.

Matching

Match each of the terms in the left column to the appropriate definition in the right column.

_____ 1. Oxidizing agent   **A.** A form of energy measured in joules

_____ 2. Ceiling jet   **B.** Any substance that can undergo combustion

_____ 3. Heat   **C.** The detailed study of how chemistry, fire science, and the engineering disciplines of fluid mechanics and heat transfer interact to influence fire behavior

_____ 4. Plume   **D.** The transport of heat energy from one point to another caused by a temperature difference between those points

_____ 5. Fuel   **E.** The column of hot gases, flames, and smoke rising above a fire

_____ 6. Fuel package   **F.** The amount of fuel present

_____ 7. Heat transfer   **G.** A collection or array of fuel items in close proximity with one another such that flames can spread throughout the array of fuel items

_____ 8. Fire dynamics   **H.** A substance that promotes oxidation during the combustion process

_____ 9. Fuel load   **I.** A measurement of the amount of molecular activity when compared with a reference or standard

_____ 10. Temperature   **J.** A relatively thin layer of flowing hot gases that develops under a horizontal surface as a result of plume impingement and the flowing gas being forced to move horizontally

Multiple Choice

Read each item carefully, and then select the best response.

_____ 1. What is a self-sustaining event that continues to develop fuel vapors and sustain flames even after the removal of the ignition source?
   A. Exothermic reaction
   B. Uninhibited chemical chain reaction
   C. Pyrolysis
   D. Combustion

_____ 2. What is a process in which something is heated, causing the material to decay and produce fire gases?
   A. Exothermic reaction
   B. Uninhibited chemical chain reaction
   C. Pyrolysis
   D. Combustion

_____ 3. What are reactions that result in the release of energy in the form of heat?
   A. Exothermic reaction
   B. Uninhibited chemical chain reaction
   C. Pyrolysis
   D. Combustion

_____ 4. What is the process of producing ignitable vapors from a liquid?
   A. Thermal decomposition
   B. Heat flux
   C. Conduction
   D. Vaporization

_____ 5. What is heat transfer by circulation within a medium such as a gas or a liquid?
   A. Heat capacity
   B. Convection
   C. Radiation
   D. Heat release rate

_____ 6. What is a chemical process of oxidation that occurs at a rate fast enough to produce heat and usually light in the form of either a glow or flames?
   A. Exothermic reaction
   B. Uninhibited chemical chain reaction
   C. Pyrolysis
   D. Combustion

_____ 7. What is the combined process of emission, transmission, and absorption of energy traveling by electromagnetic wave propagation between a region of higher temperature and a region of lower temperature?
   A. Heat capacity
   B. Convection
   C. Radiation
   D. Heat release rate

_____ 8. Which is the measurement of the rate of heat transfer to a surface?
   A. Exothermic reaction
   B. Uninhibited chemical chain reaction
   C. Pyrolysis
   D. Heat flux

_____ 9. What is the measure of energy required to raise the temperature of an object one degree of a unit mass (J/kg K)?
   A. Heat capacity
   B. Convection
   C. Radiation
   D. Conduction

_____ 10. What is the heat transfer to another body or within a body by direct contact called?
   A. Heat capacity
   B. Convection
   C. Radiation
   D. Conduction

_____ 11. What is the measure of heat that could travel across an area with a temperature gradient, expressed as one degree per unit of length (W/m K)?
   A. Heat capacity
   B. Thermal conductivity
   C. Thermal decomposition
   D. Thermal inertia

_____ 12. What are those properties of the material that characterize its rate of surface temperature rise when exposed to heat? It is the product of the thermal conductivity (κ), the density (ρ), and heat capacity (c).
   A. Heat capacity
   B. Thermal conductivity
   C. Thermal decomposition
   D. Thermal inertia

_____ 13. What energy is released by the individual fuels being consumed, measured in either watts or kilowatts?
   A. Heat capacity
   B. Heat release rate
   C. Radiation
   D. Heat flux

_____ 14. The lowest temperature at which a gas–air mixture will ignite in the absence of an ignition source is called the
   A. Flash point
   B. Fire point
   C. Autoignition temperature (AIT)
   D. Thermal runaway

_____ 15. What is the lowest temperature at which a liquid produces a flammable vapor?
   A. Flash point
   B. Fire point
   C. Autoignition temperature (AIT)
   D. Thermal runaway

Fill-in

Read each item carefully and then complete the statement by filling in the missing word(s).

1. Even though some materials are capable of self-heating, a condition of _____ _____ needs to occur for self-ignition to occur.

2. A material that is not combustible but can increase the rate of combustion or produce spontaneous combustion when combined with other substances is considered to be a(n) _____ _____.

3. Rates of flame spread are dependent on not only the individual fuel properties but also on the _____ and _____ of the fuel surfaces.

4. Flame spread may be the result of _____ or _____ materials from a fuel package.

5. During the ignition and growth phases of a compartment fire, there is a sufficient amount of air available to allow the fire to continue to burn. When the size of the fire is controlled by how much fuel is burning, this is referred to as _____-_____ _____.

6. If the rate of combustion begins to exceed the amount of air flow into the compartment, the fire transitions from fuel controlled to _____ _____.

7. One of the major determining factors in whether flashover will occur is the presence of a hot gas layer with sufficient energy to radiate downward to involve exposed _____ _____.

8. Full room involvement can result in the production of low patterns, _____ _____, and holes in the floor.

9. A fire in a compartment will have only one _____ _____ even if there are multiple ventilation openings.

10. If a fire has a limited amount of air for combustion, a(n) _____ in the amount of visible products of combustion, such as soot, smoke, and carbon monoxide, will occur.

Vocabulary

Define the following terms using the space provided.

1. Combustion:

2. Conduction:

3. Convection:

4. Flash point:

5. Flashover:

6. Oxidizing agent:

7. Pyrolysis:

8. Radiation:

9. Thermal inertia:

10. Vaporization:

Short Answer

Complete this section with short written answers, using the space provided.

1. Identify the most common fuels a fire investigator will encounter.

2. Explain the fourth component of the fire tetrahedron, an uninhibited chain reaction.

3. What is the "stoichiometric ratio"?

4. Discuss the use of smoke color and density as an indicator of material burning.

5. Describe "gas plumes" and "ceiling jet."

Fire Alarms

The following case scenario will give you an opportunity to explore the concerns associated with fire investigation. Read the scenario and then answer each question in detail.

You are investigating a compartment fire in a five-story residential high rise. The fire was confined to a single apartment, with smoke damage limited to the floor of origin. The initial engine company reported thick smoke in the apartment, but with the exception of a chair and some curtains, there was very little actual fire and minimal fire damage to the furniture and other contents.

1. What questions would you ask the tip man concerning this fire?

2. What could explain the lack of damage to the majority of the room contents?

Additional Activity

1. Using available resources, create a list of five common household chemicals that are oxidizing agents.

   _____
   _____
   _____
   _____
   _____
   _____
   _____
   _____

2. Using available resources, create a list of five common industrial oxidizers used in or transported through your jurisdiction.

   _____
   _____
   _____
   _____
   _____
   _____
   _____
   _____

Fire Patterns

Workbook Activities

The following activities have been designed to help you. Your instructor may require you to complete some or all of these activities as a regular part of your investigator training program. You are encouraged to complete any activity that your instructor does not assign as a way to enhance your learning in the classroom.

Chapter Review

The following exercises provide an opportunity to refresh your knowledge of this chapter.

Matching

Match each of the terms in the left column to the appropriate definition in the right column.

_____ 1. Char
_____ 2. Fire pattern
_____ 3. Melting
_____ 4. Crazing
_____ 5. Fire effects
_____ 6. Beveling
_____ 7. Oxidation
_____ 8. Fire pattern analysis
_____ 9. Calcination
_____ 10. Spalling

A. Basic chemical process associated with combustion
B. Observable or measurable changes in or on a material as a result of exposure to fire
C. Occurs in plaster or gypsum wall surfaces when the chemically bound water is driven out of the gypsum by the heat of the fire
D. Visible or measurable physical changes or identifiable shapes formed by a fire effect or group of fire effects
E. The process of identifying and interpreting fire patterns
F. Chipping or pitting of concrete or masonry surfaces
G. A physical change from solid to liquid caused by exposure to heat
H. Carbonaceous material that has been burned or pyrolyzed and has a blackened appearance
I. Cracks in glass produced by rapid cooling
J. An indicator of fire direction on wood wall studs

Multiple Choice

Read each item carefully and then select the best response.

_____ 1. The collection of fire scene data requires the recognition and identification of
   A. Proper tools and equipment
   B. Criminal acts
   C. Fire effects and fire patterns
   D. Usual suspects

CHAPTER 4

_____ 2. Melting, distortion, and other fire effects sustained to materials may be very useful to an investigator estimating
   A. Temperature
   B. Fuel load
   C. Fuel type
   D. Hydrocarbons

_____ 3. Depth of char can be used to assess
   A. Duration of a fire
   B. Use of ignitable liquids
   C. Ignition source
   D. Fire movement

_____ 4. Rapid heating can cause moisture in concrete to heat, expand, and crack the surface, called
   A. Oxidizing
   B. Spalling
   C. Laminating
   D. Boiling

_____ 5. A "clean burn" area can guide investigators in determining
   A. Fire spread
   B. Fuel
   C. Oxidation rate
   D. Distortion or deformity

_____ 6. Glass fragments found to be free of soot deposits of smoke condensates usually indicate
   A. Use of explosives
   B. Breaking and entering
   C. Fire-resistant glass
   D. Early failure of the glass

_____ 7. Incandescent light bulbs, 25 W or greater, may be used to determine
   A. Use of flammable liquids
   B. Use of combustible liquids
   C. Fire travel
   D. Calcinations

_____ 8. Factors that influence fire patterns include
   A. Ventilation and airflow
   B. Ignition factors
   C. Fuel loads
   D. All of the above

_____ 9. Some surfaces exposed to fire may have little or no damage due to ventilation effects and
   A. Calcinations
   B. Location of furnishings
   C. Spalling
   D. Rainbow effect

_____ 10. Penetrations in floors may be the result of ignitable liquids, smoldering items, effects of ventilation, or
   A. Radiant heat
   B. Crazing
   C. Beveling
   D. Char

Fill-in
Read each item carefully and then complete the statement by filling in the missing word(s).

1. The interpretation of _____ _____ has traditionally been one of the primary processes used in fire investigation.

2. Temperatures in most structure fires seldom remain above _____ °F for an extended period of time.

3. _____ _____ may be produced between melted and unmelted portions of a material and may be useful patterns to an investigator.

4. When exposed to heat, most materials begin to _____ and change shape.

5. When exposed to temperatures in excess of 1000°F, _____ _____ begin to buckle and bend.

6. Pressures developed by structure fires are generally _____ to cause window panes to be blown out.

7. Laboratory tests have revealed that the intensity and time of exposure both play roles in the loss of tensile strength in _____ _____.

8. _____ _____ is caused by an object blocking the travel of radiated heat, convected heat, or direct flame to a surface.

Vocabulary
Define the following terms using the space provided.

1. Beveling:

2. Calcination:

3. Crazing:

4. Fire effects:

5. Heat shadowing:

6. Rainbow effect:

7. Spalling:

Short Answer
Complete this section with short written answers, using the space provided.

1. Identify seven factors that may affect the rate at which wood may char.

2. Identify five limitations in the use of changes in color of materials as a source of information.

3. Explain the difference between steels melting and steels deforming.

4. What is the "rainbow effect" and why is it *not* a reliable indication of the presence of an ignitable liquid?

5. How may fire-suppression activities create or change fire patterns?

Fire Alarms

The following case scenario will give you an opportunity to explore the concerns associated with fire investigation. Read the scenario and then answer the question in detail.

You have been called to investigate a structure fire in a residential dwelling that has received considerable damage. During your initial examination you note a room that appears to have been fully involved with a hole in the floor and extensive fire damage to the room below. You must determine whether the fire began in the lower floor or started in the upper room.

1. What clues may help you in your investigation?

   _____
   _____
   _____
   _____
   _____
   _____
   _____
   _____
   _____
   _____

Building Systems

Workbook Activities

The following activities have been designed to help you. Your instructor may require you to complete some or all activities as a regular part of your investigator training program. You are encouraged to complete any activity your instructor does not assign as a way to enhance your learning in the classroom.

Chapter Review

The following exercises provide an opportunity to refresh your knowledge of this chapter.

Matching
Match each of the terms in the left column to the appropriate definition in the right column.

_____ 1. Type V construction
_____ 2. Dead load
_____ 3. Interstitial spaces
_____ 4. Live load
_____ 5. Type III construction
_____ 6. As-built conditions
_____ 7. Fuel load
_____ 8. Compartmentation
_____ 9. Balloon construction
_____ 10. Platform construction

A. A load that can move.
B. Subdivision into separate sections or units.
C. Wood frame construction.
D. Exterior wall studs extend from the foundation to the roof line. Fire stops will be lacking.
E. Concealed spaces in buildings where fires may develop and grow undetected.
F. Constant and immobile.
G. Exterior walls are masonry or other noncombustible material.
H. Walls are placed on top of floors, forming an effective fire stop.
I. Existing building conditions that vary from original building plans.
J. The total quantity of combustible contents of a building, space, or fire area.

Multiple Choice
Read each item carefully and then select the best response.

_____ 1. Used in a wide variety of buildings, the floor, roof, and partition framing are wood assemblies and the exterior walls are noncombustible material:
   A. Ordinary construction
   B. Commercial construction
   C. Balloon construction
   D. Residential construction

CHAPTER 5

_____ 2. This type of construction allows for large spans of unsupported finish material, which may result in failure of structural sections with large frame members still standing:
   A. Plank and beam
   B. Balloon frame
   C. Heavy timber
   D. Platform frame

_____ 3. This type of construction uses structural members that are unprotected wood, the smallest dimension being 6 or 8 inches:
   A. Plank and beam
   B. Balloon frame
   C. Heavy timber
   D. Platform frame

_____ 4. These structural elements are composed of many wood planks glued together to form one solid beam. They are generally for interior use:
   A. Wood truss
   B. Wood I-beam
   C. Laminated beam
   D. Wood joist

_____ 5. These structural elements have a smaller dimension than floor joists and can burn through and fail sooner than dimensional lumber:
   A. Wood truss
   B. Wood I-beam
   C. Laminated beam
   D. Wood joist

_____ 6. These structural elements have wood members fastened together using nails, staples, or gusset plates that can fail even before wood members burn through:
   A. Wood truss
   B. Wood I-beam
   C. Laminated beam
   D. Wood joist

_____ 7. Manufactured parts put together to make a complete product and may or may not be fire resistance rated:
   A. Trusses
   B. I-beams
   C. Soffits
   D. Assemblies

_____ 8. These assemblies are among the first to fail when structural elements are exposed to fire conditions:
   A. Floor
   B. Ceiling
   C. Roof
   D. All of the above

____ 9. When approved, this structural component has a label indicating its classification and rating:
   A. Truss
   B. Fire door
   C. Stairwell
   D. Joist

____ 10. A concept in which fire is confined to the room of origin, minimizing smoke movement to other areas of a building:
   A. Balloon construction
   B. Interstitial spaces
   C. Compartmentation
   D. Assemblies

Fill-in
Read each item carefully and then complete the statement by filling in the missing word(s).

1. The fire investigator must have an understanding of _____ _____ to analyze properly fire growth and movement as well as fire patterns that remain after a fire.

2. Fires that originate in _____ _____ may develop undetected and move freely and rapidly without barriers to stop their spread.

3. A(n) _____ _____ operating at the time of a fire can facilitate the spread of smoke and fire through a structure.

4. _____ _____ are the weight of materials that are part of a building, such as the structural components and mechanical equipment.

5. Removal of any structural elements along a _____-_____ wall could create undue stress on the building, leading to potential collapse.

6. A change in use or _____ _____ can result in the introduction of fuel loads for which the building's fire protection systems were not designed.

7. _____ _____ uses a construction technique whereby the structure is built in one or more sections and then transported to the building site and assembled.

8. A fire rating on a wood frame assembly is accomplished by covering the wall with a noncombustible finish, commonly _____ _____.

Vocabulary
Define the following terms using the space provided.

1. Compartmentation:

   _____
   _____
   _____

2. Dead load:

3. Interstitial spaces:

4. Live load:

5. Mill construction:

6. Smoke barrier:

7. Soffits:

Short Answer

Complete this section with short written answers, using the space provided.

1. Identify six conditions under which a building is no longer in balance or capable of supporting its loads.

2. Identify six factors that significantly affect fire spread in compartment fires.

3. Identify three manufactured wood products common in newer residential structures.

4. Compare features of steel, masonry, and concrete construction.

5. Identify five factors that can affect the failure of floor, ceiling, and roof assemblies.

Labeling

Label the following diagram with the correct terms.

1. Typical single-family dwelling

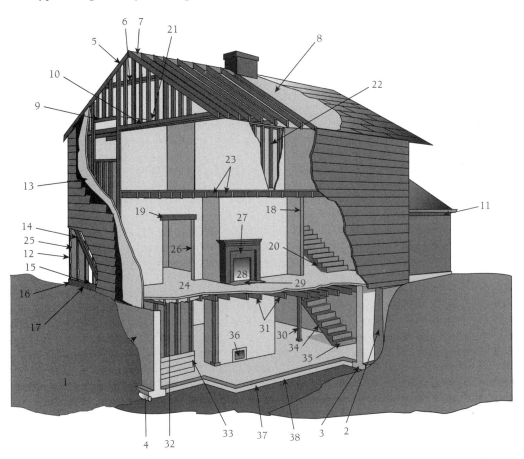

1. _____
2. _____
3. _____
4. _____
5. _____
6. _____
7. _____
8. _____
9. _____
10. _____
11. _____
12. _____
13. _____
14. _____
15. _____
16. _____
17. _____
18. _____
19. _____
20. _____
21. _____
22. _____
23. _____
24. _____
25. _____
26. _____
27. _____
28. _____
29. _____
30. _____
31. _____
32. _____
33. _____
34. _____
35. _____
36. _____
37. _____
38. _____

2. Types of building construction.

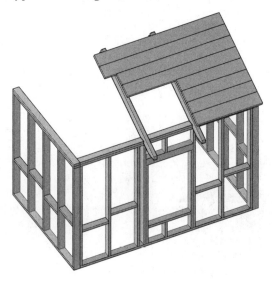

A. _____

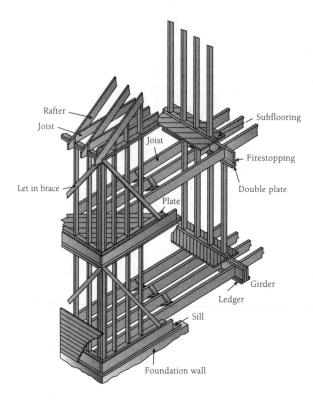

B. _____

Chapter 5 : Building Systems

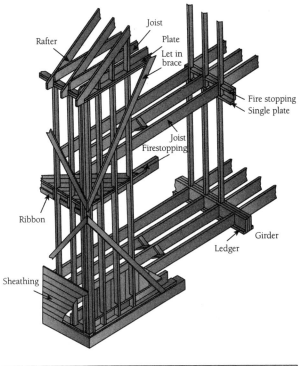

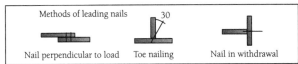

C. _____

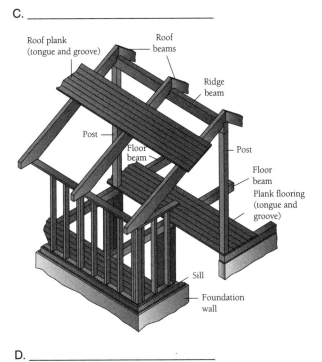

D. _____

E. _____

F. _____

Fire Alarms

The following case scenario will give you an opportunity to explore the concerns associated with fire investigation. Read the scenario and then answer each question in detail.

A new multifamily housing development is under construction in your area of jurisdiction. Many construction components delivered to the construction site are engineered wood (manufactured wood) products. Manufactured wood construction components are new to your town, as most of the existing residential structures are 45 years old or older.

1. What concerns do you have with engineered (manufactured) wood products?

_____
_____
_____
_____
_____

2. How will this new type of construction affect investigations of fires in these new structures?

Additional Activity

1. Using NIOSH Publication No. 2009-114, *Preventing Deaths and Injuries of Fire Fighters Working Above Fire-Damaged Floors* (http://www.cdc.gov/niosh/docs/wp-solutions/2009-114/default.html), describe the dangers of working above fire-damaged engineered wood floor joists and list recommendations by NIOSH to fire fighters aimed at reducing these dangers.

Electricity and Fire

Workbook Activities

The following activities have been designed to help you. Your instructor may require you to complete some or all activities as a regular part of your investigator training program. You are encouraged to complete any activity your instructor does not assign as a way to enhance your learning in the classroom.

Chapter Review

The following exercises provide an opportunity to refresh your knowledge of this chapter.

Matching

Match each of the terms in the left column to the appropriate definition in the right column.

_____ 1. Amperage

_____ 2. Service lateral

_____ 3. Overload

_____ 4. Direct current (DC)

_____ 5. Main disconnect

_____ 6. Grounding

_____ 7. Meter

_____ 8. Weatherhead

_____ 9. Alternating current (AC)

_____ 10. Service drop

A. The voltage varies in time with a sine wave at 60 cycles per second. A sine-wave current results if the load is resistive. The electrical current flows in and out from the upstream electrical source in a cycle (usually a transformer), whereas the voltage is also changing by alternating from − to + in a repeating cycle.

B. A watt-hour meter that plugs into the meter base to measure the amount of electricity consumed at a site.

C. The point where service entrance cables connect to the structure, which is designed to keep water out of the conduit that carries the wires.

D. The unit of electric current that is equivalent to a flow of one coulomb per second; one coulomb is defined as 6.24×10^{18} electrons. It is similar to the flow of water in gallons per minute.

E. Wiring coming in underground.

F. Provides the master shutoff mechanism for the overall system power and provides high current level protection of downstream overcurrent protection devices and wiring.

G. Operation of equipment in excess of normal, full-load rating or of a conductor in excess of rated ampacity that when it persists for a sufficient length of time could cause damage or dangerous overheating.

H. The overhead service conductors from the last pole or other aerial support to and including the splices, if any, connecting to the service-entrance conductors at the structure.

I. A conducting connection, whether intentional or accidental, between an electrical circuit or equipment and earth or to some conducting body that serves in place of the earth.

J. The voltage is a steady level that does not vary with time. The current will also be a steady level if the load is resistive.

CHAPTER 6

Multiple Choice
Read each item carefully and then select the best response.

_____ 1. What is the electrical charging of materials through physical contact and separation and the various effects that result from the electrical charges formed by this process?
   A. Static electricity
   B. Resistance heating
   C. Switch loading
   D. Overfusing

_____ 2. What is the maximum amount of current the device is capable of interrupting?
   A. Relaxation time
   B. Regular current rating
   C. Overcurrent
   D. Interrupting current rating

_____ 3. What is the analysis of the locations where electrical arcing has caused damage and the documentation of the involved electrical circuits?
   A. Arc
   B. Ampacity
   C. Arc-fault circuit interrupter (AFCI)
   D. Arc mapping

_____ 4. What is a device intended for the protection of personnel that functions to deenergize a circuit or portion thereof within an established period of time when a current to ground (possibly a human body) exceeds the values established for a Class A device?
   A. Branch circuits
   B. Ground-fault circuit interrupter (GFCI)
   C. High-resistance fault
   D. Hot legs

_____ 5. A basic law of electricity that defines the relationship between voltage, current, and resistance. If two of these three values are known, it is possible to determine the third.
   A. Interrupting current rating
   B. Ohm's law
   C. Resistive circuit
   D. Sine wave

_____ 6. What sensors are used to measure temperature by change of their resistance?
   A. Thermistors
   B. Time-current curve
   C. Voltage sniffer
   D. Switch loading

_____ 7. This occurs when current flows through a path that provides high resistance to current flow, such as a heating element or a resistive connection.
   A. Static electricity
   B. Resistance heating
   C. Switch loading
   D. Overfusing

_____ 8. What is the level of current above which the protective device will open, such as 15 A, 20 A, or 50 A?
   A. Relaxation time
   B. Regular current rating
   C. Overcurrent
   D. Interrupting current rating

_____ 9. What is a high-temperature luminous electric discharge across a gap where the conductor is missing or through a medium such as charred insulation?
   A. Arc
   B. Ampacity
   C. Arc-fault circuit interrupter (AFCI)
   D. Arc mapping

_____ 10. What are the two insulated conductors of a single-phase system (or three if three-phase power is used), sometimes referred to as L1, L2, or Line voltages?
   A. Branch circuits
   B. Ground-fault circuit interrupter (GFCI)
   C. High-resistance fault
   D. Hot legs

_____ 11. What waveform does AC voltage follow?
   A. Interrupting current rating
   B. Ohm's law
   C. Resistive circuit
   D. Sine wave

_____ 12. What is a noncontact voltage monitor that outputs a beep or turns on a light when voltage is present or nearby (within about an inch or less)?
   A. Thermistors
   B. Time-current curve
   C. Voltage sniffer
   D. Switch loading

_____ 13. What is a dangerous condition that occurs when the circuit protection (fuse or circuit breaker) rating significantly exceeds the ampacity of the conductor, leading to a condition in which increased heat can occur in the conductors?
   A. Static electricity
   B. Resistance heating
   C. Switch loading
   D. Overfusing

_____ 14. What is the amount of time for a charge to dissipate?
   A. Relaxation time
   B. Regular current rating
   C. Overcurrent
   D. Interrupting current rating

_____ 15. This is designed to protect against fires caused by arcing faults in home electrical wiring. The circuitry continuously monitors current flow.
   A. Arc
   B. Ampacity
   C. Arc-fault circuit interrupter (AFCI)
   D. Arc mapping

Fill-in
Read each item carefully and then complete the statement by filling in the missing word(s).

1. The most useful measurement in working with postfire circuits is the measurement of _____.

2. It is important to check the overcurrent protection device if an overload is suspected to ensure that the proper _____ _____ was chosen.

3. In large buildings there may be more than one _____ _____ for electrical power.

4. Without a(n) _____ _____, the energy delivered to a circuit or load may flow in an undesirable location or path.

5. _____ _____ open when exposed to heat or overcurrent that exceeds their rated trip current.

6. Analysis of the _____ inside and outside an arc-damaged panel is needed to determine whether arcing inside the panel was the source of ignition or was the result of heat impingement on the panel by the fire.

7. The AWG relates to size, which determines the _____ of the conductor.

8. A copper conductor's melting temperature may be affected by "alloying," or _____ _____.

9. An aluminum conductor's conductivity is lower than copper and must be two _____ _____ larger than copper for the same ampacity.

10. _____ outlets are used in bathrooms, kitchens, or other wet locations.

Vocabulary
Define the following terms using the space provided.

1. Arc-fault circuit interrupter (AFCI):

    _____
    _____
    _____

2. Eutectic melting:

    _____
    _____
    _____

3. Hot legs:

    _____
    _____
    _____

4. Overfusing:

5. Resistance heating:

6. Resistive circuit:

7. Sine wave:

8. Thermistors:

9. Voltage sniffer:

10. Weatherhead:

Short Answer

Complete this section with short written answers, using the space provided.

1. Discuss the comparisons between hydraulics and electricity.

    _____
    _____
    _____
    _____
    _____
    _____
    _____

2. Identify the three most commonly used electric conductors.

    _____
    _____
    _____
    _____
    _____
    _____
    _____

3. Identify the types of electrical insulation, and list their individual drawbacks.

    _____
    _____
    _____
    _____
    _____
    _____
    _____

4. List four conditions that must exist for ignition from an electrical source to occur.

    _____
    _____
    _____
    _____

5. Identify six methods that may generate sufficient heat for ignition from an electrical source.

    _____
    _____
    _____
    _____
    _____

Labeling

Label the following diagram with the correct terms.

1. Triplex overhead service drop.

A. _____
B. _____
C. _____
D. _____

Fire Alarms

The following case scenario will give you an opportunity to explore the concerns associated with fire investigation. Read the scenario and then answer each question in detail.

You are investigating an apartment bedroom fire that you suspect to be electrical in nature. Your initial observations suggest the origin of the fire was a bed stand beside the bed. The occupant had an 18-gauge extension cord running to the bed stand with five devices plugged into it.

1. How can you determine if the maximum amperage of the electrical circuit had been exceeded?

_____
_____
_____
_____
_____

2. What information will you need to determine if the conditions existed for ignition to occur from your suspect electrical source?

_____
_____
_____
_____
_____

Additional Activity

1. Using Ohm's law, calculate the following:

   A. Voltage if the current is 10 amps and the resistance is 12 ohms
   _____

   B. Voltage if the current is 25 amps and the resistance is 10 ohms
   _____

   C. Voltage if the current is 50 amps and the resistance is 8.8 ohms
   _____

   D. Current if the voltage is 110 volts and the resistance is 10 ohms
   _____

   E. Current if the voltage is 220 volts and the resistance is 20 ohms
   _____

   F. Current if the voltage is 440 volts and the resistance is 5 ohms
   _____

   G. Resistance if the voltage is 110 volts and the current is 10 amps
   _____

   H. Resistance if the voltage is 220 volts and the current is 5 amps
   _____

   I. Resistance if the voltage is 440 volts and the current is 40 amps
   _____

Fuel Gas Systems

Workbook Activities

The following activities have been designed to help you. Your instructor may require you to complete some or all activities as a regular part of your investigator training program. You are encouraged to complete any activity your instructor does not assign as a way to enhance your learning in the classroom.

Chapter Review

The following exercises provide an opportunity to refresh your knowledge of this chapter.

Matching

Match each of the terms in the left column to the appropriate definition in the right column.

_____ 1. Commercial propane

_____ 2. Lower explosive limit (LEL)

_____ 3. Pressure relief valve

_____ 4. Natural gas

_____ 5. Pressure gauge

_____ 6. Backflow valve

_____ 7. Gas burner

_____ 8. Liquefied petroleum gas (LPG) (propane)

_____ 9. Fusible plugs

_____ 10. Container appurtenances

A. Petroleum gases condensed to a liquid state with moderate pressure and normal temperatures to allow for more efficient distribution.

B. Devices connected to the openings in tanks and other containers—such as pressure relief devices, control valves, and gauges.

C. Derived from the refining of petroleum, a liquefied gas composed of 95 percent propane and propylene and 5 percent other gases.

D. A valve designed to open at a specific pressure, usually around 250 psi (1724 kPa). It is generally placed in the container where it releases the vapor.

E. Thermally activated devices that open and vent the contents of a container.

F. A device that allows fuel gases and air to properly mix and produce a flame.

G. A type of container appurtenance depicting the internal pressure of a tank. The gauges are connected directly to the tank or sometimes through the valve. Pressure gauges do not indicate the quantity of liquid propane in the tank.

H. A naturally occurring largely hydrocarbon gas product recovered by drilling wells into underground pockets, often in association with crude petroleum.

I. The lowest concentrations of fuel in a specified oxidant; also known as the lower flammable limits (LFL).

J. Prevents gas from reentering a container or distribution system.

CHAPTER 7

Multiple Choice
Read each item carefully and then select the best response.

_____ 1. What building system can fail, allowing fuel to come into contact with an ignition source and also adding to the spread and growth of a fire?
   A. Electrical system
   B. Fuel gas system
   C. Water system
   D. Communication system

_____ 2. Fuel that has escaped its container or delivery system is often referred to as what?
   A. Fugitive gas
   B. Truant gas
   C. Bad gas
   D. Fuel gas

_____ 3. What compound is most often used as an odorant in natural gas?
   A. Acetylene
   B. Water
   C. Propylene
   D. Butyl mercaptan

_____ 4. Per NFPA 58, *Liquefied Petroleum Gas Code*, the odorant must be detectable when the fuel gas is at what concentration?
   A. Not less than one-tenth of its lower explosive limit (LEL)
   B. Not less than one-fifth of its lower explosive limit (LEL)
   C. Not less than 1 percent of its lower explosive limit (LEL)
   D. There is no set concentration

_____ 5. An LP cylinder will have a water capacity of how many lbs?
   A. Greater than 1000 lb
   B. 60 lbs
   C. 1000 lb or less
   D. There is no set poundage

_____ 6. Which of the following is classified as a container appurtenance?
   A. Pressure relief device
   B. Control valve
   C. Gauge
   D. All of the above

_____ 7. What devices are pressure activated to prevent pressure from exceeding a predetermined maximum pressure to avoid rupture of a container?
   A. Pressure relief valve
   B. Fusible plug
   C. Backflow valve
   D. Pressure gauge

8. What devices are thermally activated devices that open and vent the contents of a container?
   A. Pressure relief valve
   B. Fusible plug
   C. Backflow valve
   D. Pressure gauge

9. What type of valves monitors the flow from the container and activates if the flow exceeds a set amount?
   A. Pressure relief valves
   B. Backflow valves
   C. Excess-flow check valves
   D. Main valves

10. What devices are attached through the gas supply system to reduce the pressure so they can be used by the appliance or utilization equipment?
    A. Fusible plug
    B. Backflow valve
    C. Excess-flow check valves
    D. Pressure regulators

11. Statistics indicate corrosion may cause as many as what percentage of all known gas leaks?
    A. 10
    B. 30
    C. 40
    D. 60

12. What device is used in gas detector surveys and may also detect the presence of other gases?
    A. Combustible gas indicator
    B. Bar hole
    C. Variable gauge
    D. Vaporizer

Fill-in

Read each item carefully and then complete the statement by filling in the missing word(s).

1. Pilot lights and open flames from the appliances served by _____ _____ often function as ignition sources.

2. _____ _____ can serve as both the first fuel (source of the ignition) and the fuel for the fire.

3. Containers are governed by U.S. Department of Transportation regulations and are required to have pressure relief valves or _____ _____.

4. _____ are frequently used when there is a demand for large quantities of propane, such as for industrial uses or in cold weather environments.

5. The most common types of pressure regulators are the _____ _____, which has a spring that is set to control the pressure, and the _____ _____.

6. Pilotless burners do not have pilots but are ignited by _____ _____ or resistance heating elements.

7. Each above-ground portion of a piping system is required to be electrically _____ and the system _____.

8. Exhaust _____ is required to prevent buildup of products of combustion inside the building.

9. As with all parts of a fire investigation, the analysis of the fuel gas system and each component of that system should be done in a _____ _____ to ensure thoroughness.

10. The amount of gas that flows from an unlit _____ _____ is generally not sufficient to cause an explosion or fuel fire unless the gas is somehow confined to a space.

Vocabulary

Define the following terms using the space provided.

1. Backflow valve:

2. Container appurtenances:

3. Fixed level gauge

4. Lower explosive limit (LEL):

5. Vaporizers:

6. Variable gauge:

Short Answer

Complete this section with short written answers, using the space provided.

1. List the differences between natural gas and propane.

   _____
   _____
   _____
   _____
   _____
   _____

2. Describe the differences between natural gas supply systems and propane (LP) supply systems.

   _____
   _____
   _____
   _____
   _____
   _____

3. Describe the similarities and differences between LP cylinders and LP tanks.

   _____
   _____
   _____
   _____
   _____
   _____

4. Describe the similarities and differences between pressure relief valves and fusible plugs.

   _____
   _____
   _____
   _____
   _____

Chapter 7 : Fuel Gas Systems 51

5. Identify two types of pressure regulators. What safeguards should be taken during the installation of these gas system devices?

_____
_____
_____
_____
_____
_____

Labeling

1. Water heater components.

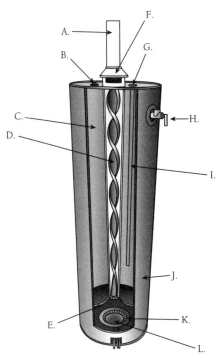

A. _____
B. _____
C. _____
D. _____
E. _____
F. _____
G. _____
H. _____
I. _____
J. _____
K. _____
L. _____

2. Furnace components.

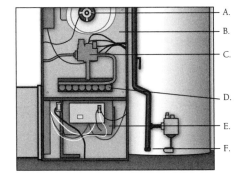

A. _____
B. _____
C. _____
D. _____
E. _____
F. _____

Fire Alarms

The following case scenario will give you an opportunity to explore the concerns associated with fire investigation. Read the scenario and then answer each question in detail.

You are investigating what appears to be an explosion with fire in the basement of a neighborhood restaurant. The restaurant is located on an intersection, with a gasoline station across the street. The restaurant owner states the gas station fuel tanks have been leaking gasoline for quite some time and the fumes have migrated into his basement and caused the explosion. The restaurant is heated by natural gas. The furnace and natural gas water heater are in the basement. The kitchen stoves and ovens also burn natural gas; they are located on the first floor.

1. Where should your investigation begin?

   _____
   _____
   _____
   _____
   _____

2. What outside assistance may you need for this investigation?

   _____
   _____
   _____
   _____
   _____

Additional Activity

1. Using Table 7-1, page 111 in the textbook, answer the following questions:

   Your combustible gas indicator has detected a natural gas leak at the top of a water heater. The water heater has an operating pilot light at the base. Your properly calibrated CGI is reading 1.5% at the leak.

   A. Is the atmosphere at the gas leak within the flammable range of natural gas?

   _____
   _____
   _____

   B. Will the natural gas have a tendency to drop down to the pilot light and ignite, or will it rise to the ceiling area and seek an ignition source there?

   _____
   _____
   _____

Fire-Related Human Behavior

Workbook Activities

The following activities have been designed to help you. Your instructor may require you to complete some or all activities as a regular part of your investigator training program. You are encouraged to complete any activity your instructor does not assign as a way to enhance your learning in the classroom.

Chapter Review

The following exercises provide an opportunity to refresh your knowledge of this chapter.

Matching
Match each of the terms in the left column to the appropriate definition in the right column.

_____ 1. Cognitive comprehension limitations
_____ 2. Small groups
_____ 3. Large groups
_____ 4. Physical limitations
_____ 5. Fire incidents
_____ 6. Child fire-setters
_____ 7. Adolescent fire-setters
_____ 8. Formalized group
_____ 9. Recall notice
_____ 10. Statement of danger

A. The very young and very old are most susceptible to these limitations
B. Members not in a leadership role tend to delay their response until a leader responds to the situation
C. Notifies consumers of a product defect that was identified after the product was released
D. Can cause a significant delay in an individual's response time
E. Have a tendency not to respond within appropriate time frames
F. Frequently occur as a result of an act of omission by one or more individuals
G. Aid in response time of the individuals associated with it
H. Usually symptomatic of stress, anxiety, anger, or other psychological or emotional problem
I. Identifies the nature and extent of the danger and the gravity of the risk of injury
J. Two to 6 years old, curious, and set fires out of sight of adults

Multiple Choice
Read each item carefully and then select the best response.

_____ 1. These limitations can affect an individual's ability to recognize and react appropriately to a fire:
   A. Cognitive comprehension
   B. Physical
   C. Familiarity
   D. Training

CHAPTER 8

_____ **2.** These limitations can affect an individual's ability to take appropriate actions before and during a fire:
  A. Cognitive comprehension
  B. Physical
  C. Familiarity
  D. Training

_____ **3.** These limitations can affect an individual's ability to escape a structure or setting:
  A. Cognitive comprehension
  B. Physical
  C. Familiarity
  D. Training

_____ **4.** An individual's response to a threat or purported threat is tempered by
  A. Demographics of the group
  B. Size of the group
  C. Age of the group
  D. Permanence of the group

_____ **5.** During an emergency, occupants unfamiliar with a building will tend to exit
  A. By a window
  B. Through the closest door
  C. Unpredictably
  D. Through the door they entered

_____ **6.** If an alarm system has had a number of false alarms, occupants tend to
  A. Leave immediately
  B. Call the alarm company to verify an emergency exists
  C. Delay their response
  D. Panic

_____ **7.** A positive effect of fire suppression systems is that occupants
  A. Have more time to react
  B. Tend to fight the fire
  C. Can remain at their job
  D. Have increased visibility

_____ **8.** A negative effect of discharged fire suppression systems is that
  A. Occupants tend to fight the fire
  B. Occupants tend to remain at their job
  C. Occupants' visibility may be decreased
  D. Occupants tend to watch the fire continue to develop

_____ 9. During the lifetime of most equipment, there is a prescribed maintenance and cleaning schedule that is provided by the manufacturer to prevent
   A. Misuse
   B. Malfunction
   C. Recalls
   D. Complaints

_____ 10. Fires set by juvenile fire-setters are often set in and around the home or
   A. Church
   B. Automobile
   C. Vacant lot
   D. Educational setting

Fill-in
Read each item carefully and then complete the statement by filling in the missing word(s).

1. An individual's actions are shaped by _____ _____, including his or her physical limitations, limitations of cognitive comprehension, and knowledge of the physical setting.

2. An individual's _____ with the setting can make escape more likely, although physical limitations and cognitive impairments can minimize this advantage.

3. An individual's response to a threat is tempered by the _____ of the group with which the individual is associated.

4. Research has indicated that the degree of familiarity among the individuals in a group affects _____ _____.

5. A group's roles and norms in terms of gender, social class, occupation, and so forth can affect its response to _____.

6. Operating procedures for equipment or appliances are designed to ensure _____.

7. The investigator should note any housekeeping _____ that may have contributed to a fire or explosion.

8. A _____ _____ _____ identifies the nature and extent of the danger and the gravity of the risk of injury.

Vocabulary
Define the following terms using the space provided.

1. Adolescent fire-setters:

   _____
   _____
   _____

2. Child fire-setters:

   _____
   _____
   _____

3. Juvenile fire-setters:

_____

_____

_____

4. Statement of the danger:

_____

_____

_____

Short Answer

Complete this section with short written answers, using the space provided.

1. Describe how an individual's behavior may be affected by his or her familiarity with the physical setting of the fire or explosion.

_____

_____

_____

_____

_____

_____

2. Describe the three age groups of juvenile fire-setters and the reasons they are drawn to setting fires.

_____

_____

_____

_____

_____

_____

3. Identify and define the alert words used to draw a user's attention to warnings on product labels.

4. Describe how improper maintenance of equipment can lead to a fire or explosion.

5. Identify the types of information an investigator should be seeking during a personal interview with a fire survivor.

Fire Alarms

The following case scenario will give you an opportunity to explore the concerns associated with fire investigation. Read the scenario and then answer each question in detail.

You have begun an investigation into a suspicious fire at a child-care foster home. The seven occupants are five children ranging in age from 6 to 16 and two adults.

1. How will you factor in occupant age as you conduct your initial investigation?

   _____
   _____
   _____
   _____
   _____

2. All occupants escaped safely, although the fire occurred at 2:00 AM and the structure was 60 percent involved in fire. What could account for the 100 percent survival rate of occupants?

   _____
   _____
   _____
   _____
   _____

Legal Considerations

Workbook Activities

The following activities have been designed to help you. Your instructor may require you to complete some or all activities as a regular part of your investigator training program. You are encouraged to complete any activity your instructor does not assign as a way to enhance your learning in the classroom.

Chapter Review

The following exercises provide an opportunity to refresh your knowledge of this chapter.

Matching

Match each of the terms in the left column to the appropriate definition in the right column.

_____ 1. Sufficient probable cause

_____ 2. Fifth Amendment

_____ 3. Real evidence

_____ 4. Demonstrative evidence

_____ 5. Sixth Amendment

_____ 6. Chain of custody

_____ 7. Miranda Rule

_____ 8. Documentary evidence

_____ 9. Interested party

_____ 10. Fourth Amendment

A. Right to counsel

B. Prohibits unreasonable searches and seizures

C. Requires that a witness being interrogated or interviewed in a custodial setting receive certain warnings before the statement can legally be taken and used as evidence

D. Reasonably trustworthy information that would lead a reasonable person to believe the suspect had committed a crime

E. Physical items that may be produced in court for inspection by the judge and jury

F. Tangible evidence such as a photograph, diagram, or chart that is used to demonstrate an issue relevant to the case

G. Any person, entity, or organization, including their representatives, with statuary obligations or whose rights or interests may be affected by the investigation

H. Protection against self-incrimination

I. A record of those who had the evidence in their custody and how it was moved

J. Any type of written record or document that is relevant to the case

CHAPTER 9

Multiple Choice

Read each item carefully and then select the best response.

_____ 1. A voluntary written statement of fact or opinion is a(n)
   A. Deposition
   B. Interrogatory
   C. Affidavit
   D. Motion

_____ 2. A set of questions that is served by one party involved in litigation to another involved party:
   A. Deposition
   B. Interrogatory
   C. Affidavit
   D. Motion

_____ 3. An oral testimony made under oath is a(n)
   A. Deposition
   B. Interrogatory
   C. Affidavit
   D. Motion

_____ 4. A person who testifies about matters within his or her own firsthand knowledge:
   A. Expert witness
   B. Defendant
   C. Lay witness
   D. Prosecutor

_____ 5. A person sued in a civil proceeding or accused in a criminal proceeding:
   A. Expert witness
   B. Defendant
   C. Lay witness
   D. Plaintiff

_____ 6. An individual who is specially trained to collect, preserve, and transport physical evidence:
   A. Expert witness
   B. Defendant
   C. Lay witness
   D. Evidence technician

_____ 7. Evidence based on inference and not on personal observation:
   A. Documentary
   B. Demonstrative
   C. Testimonial
   D. Circumstantial

_____ 8. An inner drive or impulse that is the cause, reason, or incentive that induces or prompts a specific behavior:
   A. Motive
   B. Motion
   C. Arson
   D. Testimonial

_____ 9. Conduct that falls within the legal standard established to protect others against harm is
   A. Malfeasance
   B. Negligence
   C. Nonfeasance
   D. Malpractice

_____ 10. Indicators for an investigator to consider that might lead to the inference that a fire was incendiary:
   A. Multiple fires
   B. Trailers
   C. Presence of ignitable liquids
   D. All of the above

Fill-in

Read each item carefully and then complete the statement by filling in the missing word(s).

1. An investigator who enters a scene to conduct an investigation or gather evidence without legal authority or the necessary consent might face _____ _____.

2. Under the Fourth Amendment of the U.S. Constitution, every entry onto a fire scene must be _____.

3. A _____ _____ is perhaps the most convenient method of entry to ensure compliance with the Fourth Amendment.

4. When an individual has been given a Miranda warning and agrees to make a statement, a written _____ _____ _____ should be signed and witnessed.

5. An officer must make an arrest based on _____ _____.

6. An investigator can face claims of _____ when other parties later find they are unable to investigate independently the evidence from the fire scene because that evidence is no longer available as it existed immediately after the fire.

7. To be admissible at trial, evidence must be properly transported and stored and must be authenticated by establishing a _____ _____ _____.

8. _____ _____ is verbal testimony of a witness given under oath or affirmation and subject to cross-examination by the opposing party.

Vocabulary

Define the following terms using the space provided.

1. Arson:

_____

_____

_____

2. Circumstantial evidence:

3. Defendant:

4. Demonstrative evidence:

5. Deposition:

6. Incendiary fire:

7. Tort:

Short Answer

Complete this section with short written answers, using the space provided.

1. Identify and describe the four forms of evidence used at trial.

2. Identify five factors a judge may consider when evaluating an expert's testimony.

3. Identify and describe nine basic components of an expert's qualifications.

4. Describe four examples of circumstantial evidence that may lead an investigator to believe a fire was intentional.

5. List 15 potential offenses that may be relevant in a fire investigation.

___

Fire Alarms

The following case scenario will give you an opportunity to explore the concerns associated with fire investigation. Read the scenario and then answer the question in detail.

You are investigating a residential structure fire with a fatality. The fire appears to have started in the kitchen, at a coffee maker. The coffee maker model had recently been recalled for a defective heating unit.

1. What investigative tasks may be required by you?

___

Safety

Workbook Activities

The following activities have been designed to help you. Your instructor may require you to complete some or all activities as a regular part of your investigator training program. You are encouraged to complete any activity your instructor does not assign as a way to enhance your learning in the classroom.

Chapter Review

The following exercises provide an opportunity to refresh your knowledge of this chapter.

Matching

Match each of the terms in the left column to the appropriate definition in the right column.

_____ 1. Cross-contamination
_____ 2. Carbon monoxide
_____ 3. Eye protection
_____ 4. Impact load
_____ 5. Parapet wall
_____ 6. Collapse zone
_____ 7. Secondary device
_____ 8. APRs
_____ 9. OSHA
_____ 10. NFPA 1500

A. Must be identified by markers or specialized tape
B. Portion of a wall that extends above the level of the roof
C. Can only be used when there is a sufficient amount of oxygen in the atmosphere
D. A sudden added load that may impact structural stability
E. A device meant to deploy or explode after the initial incident
F. Occupational Safety and Health Administration; requires all employees have a safe workplace
G. Standard on Fire Department Safety and Health
H. Exposes the investigator, coworkers, and their families to the toxins from a fire scene
I. Crucial because many toxic substances can be absorbed through the sclera
J. Joins with hemoglobin 250 times more readily than oxygen does, preventing the body's cells from receiving oxygen

Multiple Choice

Read each item carefully and then select the best response.

_____ 1. Fire investigators must follow the safety principles that are used during
   A. Firefighting operations
   B. Crowd control
   C. Chemistry labs
   D. Station duties

CHAPTER 10

_____ 2. It is extremely rare for a fire investigation to be conducted
   A. In adverse weather
   B. Under emergency conditions
   C. In dangerous conditions
   D. In unhealthy conditions

_____ 3. It is the responsibility of the investigator to ensure all personnel who assist with the investigation are made aware of
   A. All evidence found
   B. ID of all suspects
   C. Method of extinguishment
   D. All hazards and precautions

_____ 4. One of the first tasks the investigator should conduct at a fire scene is the assessment of
   A. Suspects
   B. Witnesses
   C. Hazards and risks
   D. Equipment on scene

_____ 5. Biological hazards on a fire scene include
   A. Building instability
   B. Gas, electric, water
   C. Bacteria, virus, insects, plants
   D. Cyanide, carbon monoxide, phosgene

_____ 6. Risks at a fire scene may be controlled by
   A. Engineering controls, administrative controls, PPE
   B. Police, National Guard, SWAT
   C. Burnbacks, trench cuts
   D. Plugging and diking

_____ 7. NFPA 1971 and NFPA 1977 are two standards that outline
   A. Investigative technique
   B. Firefighting PPE requirements
   C. Hazard intervention
   D. Evidence preservation

_____ 8. Fire investigator respiratory protection can be accomplished using
   A. ASCBA
   B. APRs
   C. Engineering controls
   D. All of the above

9. Before entering any fire scene, the investigator must assess the stability of the
   A. IC
   B. Crowd
   C. Structure
   D. Suppression efforts

10. A building has two different types of loads, the dead load and the
    A. Live load
    B. Static load
    C. Unintended load
    D. Utility load

Fill-in
Read each item carefully and then complete the statement by filling in the missing word(s).

1. A freestanding wall that is no longer supported by other structural members should never be trusted without _____.

2. If the stability of a structure is a concern, the investigative team should make the work area safer by _____ _____ _____.

3. An investigator may _____ make the assumption that a utility has been disconnected by the suppression crew.

4. A commonly used terrorist tactic is to leave _____ devices.

5. Any item worn or used in an investigation has a risk of _____ after being used in hazardous environments.

6. The fire investigator has a responsibility to contact the _____ _____ when he or she arrives at the scene.

7. When fire suppression crews clear the scene, there must be a _____-_____-_____ transfer of command from the incident commander to the fire investigator in charge of the investigation.

8. The person who commits arson does not consider the _____ of others before his or her own interests.

Vocabulary
Define the following terms using the space provided.

1. Collapse zone:

2. Control of Hazardous Energy (Lockout/Tagout) standard:

3. HAZWOPER (HAzardous Waste OPerations and Emergency Response):

_____
_____
_____

4. Impact load:

_____
_____
_____

5. Permit-Required Confined Space standard:

_____
_____
_____

Short Answer

Complete this section with short written answers, using the space provided.

1. Identify six hazards at a residential fire scene that can cause harm to the health and safety of a fire investigator.

_____
_____
_____
_____

2. Identify the PPE a fire investigator might be required to wear to investigate a fire scene.

_____
_____
_____
_____

3. Identify various types of safety equipment a fire investigator may need to provide a safe work environment for his or her investigation.

_____
_____
_____
_____

4. List three types of facilities or sites an investigator may have to respond to where chemical, biological, or radioactive exposure is a threat.

5. Explain freelancing and why it is dangerous for a fire investigator.

Fire Alarms

The following case scenario will give you an opportunity to explore the concerns associated with fire investigation. Read the scenario and then answer each question in detail.

You have responded to a local hardware store fire to investigate cause. The fire was contained to the back half of the store, but part of the ceiling has collapsed. The fire is still smoldering in places, and the fire companies are completing their overhaul before they pick up and return to quarters.

1. What health hazards are you concerned with at this scene?

2. What information should you get from the departing IC?

3. What, if any, PPE would you choose to wear during your initial investigation?

_____
_____
_____
_____
_____
_____

Additional Activity

1. Calculate the additional load the following streams will add to a fire building's structural members if the water is not drained from the building:

    A. 1¾" handline flowing 175 gpm for 20 minutes
    _____

    B. 1¾" handline flowing 150 gpm for 15 minutes
    _____

    C. 2½" handline flowing 225 gpm for 35 minutes
    _____

    D. 2½" handline flowing 250 gpm for 30 minutes and 1¾" handline flowing 175 gpm for 40 minutes
    _____

Sources of Information

Workbook Activities

The following activities have been designed to help you. Your instructor may require you to complete some or all activities as a regular part of your investigator training program. You are encouraged to complete any activity your instructor does not assign as a way to enhance your learning in the classroom.

Chapter Review

The following exercises provide an opportunity to refresh your knowledge of this chapter.

Matching

Match each of the terms in the left column to the appropriate definition in the right column.

_____ 1. Municipal government

_____ 2. County government

_____ 3. Privileged communications

_____ 4. State government

_____ 5. Municipal clerk

_____ 6. Building department

_____ 7. Confidential communications

_____ 8. County assessor

_____ 9. County coroner

_____ 10. State fire marshall

A. Statements made under circumstances showing that the speaker intended the statements only for the ears of the person addressed

B. Subnational entities that function both independently and with the federal government

C. Maintains all records related to municipal licensing and municipal operations

D. Maintains records related to property and plots, including property owners, addresses, and taxable value

E. Can provide information in regard to fire incidents within the state, building inspection records, fireworks and pyrotechnics, and boiler inspections

F. Can provide information related to the identification of victims, manner and cause of death, as well as many items found either near or on the victim

G. Has records related to building, electrical, and plumbing permits and archived building blueprints and files

H. Creates and enforces laws and regulations specific to their jurisdictional boundaries

I. A subdivision within the state that consists of various townships, villages, and cities

J. Statements made by certain persons within a protected relationship such as husband–wife, priest–penitent, attorney–client

CHAPTER 11

Multiple Choice

Read each item carefully and then select the best response.

_____ 1. Maintains all records related to municipal licensing and municipal operations:
   A. Assessor
   B. Treasurer
   C. Clerk
   D. Police

_____ 2. Maintains all public records related to real estate including plot plans, maps, and taxable real property:
   A. Assessor
   B. Treasurer
   C. Clerk
   D. Police

_____ 3. Can provide public records related to names and addresses of property owners, legal descriptions of property, and the amount of paid or owed taxes on a property:
   A. Assessor
   B. Treasurer
   C. Clerk
   D. Police

_____ 4. Can provide records related to local criminal investigations:
   A. Assessor
   B. Treasurer
   C. Clerk
   D. Police

_____ 5. Responsible for recording legal documents that determine ownership of real property and maintains files of birth, death, and marriage records as well as bankruptcy documents:
   A. County recorder
   B. County clerk
   C. County treasurer
   D. County Sheriff's Department

_____ 6. Can provide both investigative and technical support for county criminal investigations and provides polygraph services, evidence collection, and retention:
   A. County recorder
   B. County clerk
   C. County treasurer
   D. County Sheriff's Department

_____ 7. Maintains public records related to civil litigation, probate records, and other documents related to county business:
   A. County recorder
   B. County clerk
   C. County treasurer
   D. County Sheriff's Department

_____ 8. Can provide information related to property owners, tax mailing addresses, legal descriptions, and the amount of either owed or paid taxes on property and also maintains all county financial records:
   A. County recorder
   B. County clerk
   C. County treasurer
   D. County Sheriff's Department

_____ 9. Source of information such as professional licenses, results of licensing exams, and regulated businesses:
   A. State Department of Vital Statistics
   B. State Department of Regulation
   C. State Department of Revenue
   D. State Department of Transportation

_____ 10. May assist with locating tax records of individuals or corporations, both past and present, as well as locating individuals through child support records:
   A. State Department of Vital Statistics
   B. State Department of Regulation
   C. State Department of Revenue
   D. State Department of Transportation

Fill-in

Read each item carefully and then complete the statement by filling in the missing word(s).

1. The availability of information to the fire investigator is governed by _____ _____.

2. The _____ _____ _____ _____ provides for public access to information held by the federal government.

3. Electronic information is gathered through the use of _____.

4. Investigators should conduct interviews _____ _____ _____ _____ to ensure witnesses are located, identified, and interviewed in a timely manner.

5. The _____ of a witness should always be established by the fire investigator.

6. During an interview, the investigator must not only record the information provided but also determine the _____ and _____ of the information.

7. Witnesses should be interviewed as soon as possible to ensure the facts are _____ remembered by the witness and not clouded by _____ with other witnesses.

8. Questions should be meaningful and designed to _____ _____ from the witness.

Vocabulary

Define the following terms using the space provided.

1. Confidential communication:

_____

_____

_____

2. Privileged communication:

Short Answer

Complete this section with short written answers, using the space provided.

1. List three types of information private sources may provide investigators that may not be available via government agencies.

2. Describe how the National Fire Protection Agency may assist an investigation of a fire or explosion.

3. Identify two private sources of information that can provide fire investigation related training.

4. Describe the types of information insurance companies can supply to fire investigators.

5. Describe four forms of information that can be gathered by a fire investigator.

Fire Alarms

The following case scenario will give you an opportunity to explore the concerns associated with fire investigation. Read the scenario and then answer each question in detail.

You have been assigned to investigate a commercial structure fire in a carpet store and warehouse. The structure has sustained a great deal of damage, and all building contents have been damaged. On initial investigation you learn the owner is experiencing financial difficulties and is behind on several bills. His business has also been very slow due to the flagging economy.

1. Develop an interview plan that includes a relevant questioning strategy for the owner and all of the business associates.

2. Establish a list of potential sources of information that may aid in your interview and investigation.

Planning and Preplanning the Investigation

Workbook Activities

The following activities have been designed to help you. Your instructor may require you to complete some or all activities as a regular part of your investigator training program. You are encouraged to complete any activity your instructor does not assign as a way to enhance your learning in the classroom.

Chapter Review

The following exercises provide an opportunity to refresh your knowledge of this chapter.

Matching

Match each of the terms in the left column to the appropriate definition in the right column.

____ 1. Hot weather	A. Provides resources when needed
____ 2. Team concept	B. Overhead hazards, foot hazards, slips and falls, electrical issues
____ 3. Hazardous materials	C. Ideal for explaining goals and making introductions
____ 4. Preinvestigation team meeting	D. May require an early morning investigation with frequent breaks
____ 5. Preplanning	E. Provides for safety and security
____ 6. Common safety issues	F. May require the use of special PPE

Multiple Choice

Read each item carefully and then select the best response.

____ 1. A person who can provide information regarding building fire alarm systems, energy systems, power supplies, or other electrical systems or components:
  A. Materials engineer or scientist
  B. Mechanical engineer
  C. Electrical engineer
  D. Chemical engineer or chemist

____ 2. A person with specialized knowledge of how materials react to different conditions, including heat and fire:
  A. Materials engineer or scientist
  B. Mechanical engineer
  C. Electrical engineer
  D. Chemical engineer or chemist

CHAPTER 12

_____ 3. A person qualified to help identify and analyze possible failure modes where chemicals are concerned:
   A. Materials engineer or scientist
   B. Mechanical engineer
   C. Electrical engineer
   D. Chemical engineer or chemist

_____ 4. A person qualified to analyze complex mechanical systems or equipment:
   A. Materials engineer or scientist
   B. Mechanical engineer
   C. Electrical engineer
   D. Chemical engineer or chemist

_____ 5. A variety of experts who can provide advice and assistance in understanding the dynamics of fire spread from the origin, the energy need for ignition, issues relating to causation, spread, fire dynamics, and in the reaction of fire protection and detection systems:
   A. Fire science and engineering
   B. Fire protection engineer
   C. Fire engineering technologist
   D. Fire engineering technician

_____ 6. A person with bachelor of science degrees in fire engineering technology, fire and safety engineering technology, or a similar discipline, or recognized equivalent, who has studied various topics related to fire science, investigation, suppression, extinguishment, fire prevention, hazardous materials, fire-related human behavior, safety and loss management, fire and safety codes and standards, and fire science research:
   A. Fire science and engineering
   B. Fire protection engineer
   C. Fire engineering technologist
   D. Fire engineering technician

_____ 7. A person who deals with the relationship of ignition sources to materials to determine what may have started the fire, knows how fire affects materials and structures, is able to assist in the analysis of how a fire detection or suppression system may have failed, and knows crucial information about building and fire codes, fire test methods, fire performance of materials, computer modeling of fires, and failure analysis:
   A. Fire science and engineering
   B. Fire protection engineer
   C. Fire engineering technologist
   D. Fire engineering technician

8. A person with an associate of science level degree in fire and safety engineering technology or similar disciplines, or recognized equivalent, who has studied various topics related to fire science, investigation, suppression, fire prevention, hazardous materials, fire-related human behavior, safety and loss management, fire and safety codes and standards, and fire science research:
   A. Fire science and engineering
   B. Fire protection engineer
   C. Fire engineering technologist
   D. Fire engineering technician

9. A person who is an expert in a field related to a specialized industry, piece of equipment, or processing system involved in an investigation:
   A. Industry expert
   B. Attorney
   C. Insurance agent
   D. Canine teams

10. A person who can offer information regarding the building and its contents before the fire, fire protection systems in the building, and the condition of those systems:
   A. Industry expert
   B. Attorney
   C. Insurance agent
   D. Canine teams

Fill-in

Read each item carefully and then complete the statement by filling in the missing word(s).

1. A fire scene investigation includes photography, sketching, evidence collection, witness interviews, and other varied tasks that require _____ _____.

2. If the investigator conducts the investigation alone, _____ _____ are still important and must be addressed.

3. Preplanning for expertise of various fields will provide _____ when needed to respond to a particular type of incident or need.

4. The _____ _____ should use special talents or training that individual team members possess in the areas of electrical, heating and air conditioning, and other engineering fields as needed.

5. Although all fires are _____, certain information must be gathered at all fire scenes during the initial assignment and response to the investigation.

6. The underlying conditions of the investigation must be obtained to respond to the incident properly with _____ _____ _____.

7. The investigator must evaluate the _____ of the resources or experts to ensure that they meet the needs of the investigation.

8. _____ with other investigators or associations is of particular benefit to identifying resources in meeting these needs and limitations.

Chapter 12 : Planning and Preplanning the Investigation

Vocabulary

1. Preinvestigation team meeting:

   _____
   _____
   _____

Short Answer
Complete this section with short written answers, using the space provided.

1. Describe the advantages of using the team concept when investigating a fire.

   _____
   _____
   _____
   _____
   _____
   _____
   _____

2. List eight items that should be included by an investigator as basic incident information.

   _____
   _____
   _____
   _____
   _____
   _____
   _____

3. Discuss the importance of scene security and identify some common scene safety issues.

   _____
   _____
   _____
   _____
   _____
   _____
   _____

4. Identify six basic functions or duties commonly performed during each investigation.

_____

_____

_____

_____

_____

_____

5. Identify PPE and equipment an investigator should have available and wear during an investigation.

_____

_____

_____

_____

_____

_____

_____

Fire Alarms

The following case scenario will give you an opportunity to explore the concerns associated with fire investigation. Read the scenario and then answer each question in detail.

Your fire department is on scene of a working chemical plant fire. The fire is contained to a lab area, which appears to be heavily damaged. You are responding to begin the fire cause investigation.

1. How will you determine the type of expert resources you may need?

   _____
   _____
   _____
   _____
   _____

2. Identify any other resources you may need for the initial and extended investigation.

   _____
   _____
   _____
   _____
   _____

Documentation of the Investigation

Workbook Activities

The following activities have been designed to help you. Your instructor may require you to complete some or all activities as a regular part of your investigator training program. You are encouraged to complete any activity your instructor does not assign as a way to enhance your learning in the classroom.

Chapter Review

The following exercises provide an opportunity to refresh your knowledge of this chapter.

Matching
Match each of the terms in the left column to the appropriate definition in the right column.

_____ 1. Cold weather
_____ 2. Pixel
_____ 3. Photography
_____ 4. Ring flash
_____ 5. Photo diagram
_____ 6. Photo painting
_____ 7. Lens
_____ 8. Bracketing
_____ 9. Mosaic photographs
_____ 10. Sequential photography

A. Taking a series of photographs with sequentially adjusted exposures
B. Use of separate lighting to increase the detail of a photograph
C. Smallest dot in an electronic image
D. Allows the observer of a photograph to understand better the totality of the view and the relationship of the subject to the overall surroundings
E. A series of photographs that encompass a large area by overlapping the start of one photograph where the previous photograph ended
F. A diagram of the investigation site, identifying the point from which each photograph was taken
G. Drains camera batteries quickly
H. Used for close-up photography
I. Gathers light and focuses image on the surface of the film
J. Provides the investigator with pictures of the scene that can be used as points of reference when writing the report

Multiple Choice
Read each item carefully and then select the best response.

_____ 1. Newer technology that provides important low-light and self-focusing features:
   A. 8 mm
   B. Beta
   C. Digital video
   D. VHS

CHAPTER 13

_____ 2. The camera used to document testimony, such as interviews of witnesses:
   A. 35 mm
   B. Video camera
   C. SLR
   D. ASA 400

_____ 3. These photographs document damage to the structure and should be taken from multiple views:
   A. Structural photographs
   B. Interior photographs
   C. Mosaic photographs
   D. Photo diagram

_____ 4. The use of a neutral UV filter on an SLR digital or film camera is recommended for any photo to be used in
   A. Mosaic photographs
   B. Sequential photographs
   C. A court proceeding
   D. Photo diagram

_____ 5. This can depict a higher percentage of the inside of a small room, but it can also distort the peripheries of the photograph, exaggerating curves in objects:
   A. Telephoto lens
   B. Fish-eye lens
   C. Filter lens
   D. 18-55 mm lens

_____ 6. The easiest light source available to the investigator is
   A. Portable lighting
   B. Flashlight
   C. Floodlight
   D. The sun

_____ 7. This specialized flash unit fits on the end of the lens and is often used when shooting a critical piece of evidence such as an arc mark or tool mark:
   A. Ring flash
   B. Lens filter
   C. Fish eye
   D. Telephoto

_____ 8. Activities an investigator should record while at the scene during his or her investigation include
   A. Conditions on arrival
   B. Suppression and overhaul activities
   C. Observers
   D. All of the above

_____ 9. Drawings that identify locations of rooms, stairs, windows, doors, and other features are called
   A. Area sketches
   B. Floor plans
   C. Exploded views
   D. Details and sections

10. A list that details the types of equipment used within a structure is known as a
   A. Sketch
   B. Floor plan
   C. Schedule
   D. Legend

Fill-in
Read each item carefully and then complete the statement by filling in the missing word(s).

1. Photographs should be taken in a predetermined manner and in accordance with _____ _____ in the fire and explosion investigation field.

2. The photographic documentation of the scene should depict _____ _____ of the scene.

3. When recording the scene, the investigator should annotate a _____ _____ _____ _____, identifying the point from which each photograph was taken, the direction of the photograph, the placement of the item, and the photo number.

4. The fire investigator should create a _____ for any drawing, indicating what the referenced symbols represent.

5. When an architect or engineer prepares a drawing, the types of materials used in construction are listed on the _____ _____.

6. In addition to visual representation of the scene, the investigator should incorporate investigative _____ _____ to supplement those items that cannot be photographed or sketched.

7. The purpose of a _____ is to communicate the observation, analysis, and conclusions made during an investigation.

8. _____ _____ are becoming an effective and comprehensive medium for exhibiting large quantities of visual material.

Vocabulary
Define the following terms using the space provided.

1. Bracketing:
   _____
   _____
   _____

2. Photo painting:
   _____
   _____
   _____

3. Ring flash:

4. Sequential photography:

Short Answer
Complete this section with short written answers, using the space provided.

1. Describe a technique effective for photographically documenting a residential, single room and contents fire scene.

2. Identify all items that should be included on a photo diagram.

3. List fire scene activities that should be documented and included with the report.

4. What information should be documented to help establish scene location?

5. Discuss the differences and similarities between diagrams and sketches.

Fire Alarms

The following case scenario will give you an opportunity to explore the concerns associated with fire investigation. Read the scenario and then answer each question in detail.

You have begun photographing and documenting an apartment fire with a fatality. It is 3:00 AM, and the structure has sustained heavy smoke damage but little structural damage.

1. What initial observations and notes should you record?

2. What concerns will you have while photographing the victim?

_____
_____
_____
_____
_____
_____

3. List all items that should be included in your scene diagram(s).

_____
_____
_____
_____
_____
_____

Physical Evidence

Workbook Activities

The following activities have been designed to help you. Your instructor may require you to complete some or all activities as a regular part of your investigator training program. You are encouraged to complete any activity your instructor does not assign as a way to enhance your learning in the classroom.

Chapter Review

The following exercises provide an opportunity to refresh your knowledge of this chapter.

Matching
Match each of the terms in the left column to the appropriate definition in the right column.

_____ 1. Physical evidence

_____ 2. Direct evidence

_____ 3. Artifact evidence

_____ 4. Demonstrative evidence

_____ 5. Circumstantial evidence

_____ 6. Fire patterns

_____ 7. Accelerant

_____ 8. Headspace

_____ 9. Cross-contamination

_____ 10. Comparative examination

A. The visual or measurable physical effects that remain after a fire

B. Facts that usually attend other facts to be proven and are drawn by logical inference from them

C. The unintentional transfer of a substance from one fire scene or location contaminated with a residue to an evidence collection site

D. Testimony of witnesses who observe acts or detect something through their five senses and surveillance equipment

E. The zone inside a sealed evidence can between the top of the fire debris and the bottom of the lid

F. Photographs, maps, X-rays, visible tests, and demonstrations

G. Generally, a function of an engineering examination performed in a laboratory

H. Any tangible item that tends to prove or disprove a particular fact or issue

I. Any fuel or oxidizer, often an ignitable liquid, used to initiate or increase the rate of growth or speed the spread of fire

J. May include remains of the first fuel ignited, the competent source of ignition, and other materials that influenced the fire's growth and development

Multiple Choice

Read each item carefully and then select the best response.

_____ 1. At a fire scene the entire scene, including the fire patterns, sources of ignition, security and fire detection equipment, and items associated with the cause of the fire, is considered
   A. Circumstantial evidence
   B. Physical evidence
   C. Direct evidence
   D. Demonstrative evidence

_____ 2. Fire patterns, the visual or measurable physical effects that remain after a fire, are
   A. Circumstantial evidence
   B. Physical evidence
   C. Direct evidence
   D. Demonstrative evidence

_____ 3. Any materials thought to be debris or rubbish that are removed from a scene should be
   A. Hosed down
   B. Hauled away
   C. Contained and marked
   D. Shoveled into the debris pile

_____ 4. The best way to protect physical evidence on a fire ground is to
   A. Remove the evidence
   B. Photograph and destroy the evidence
   C. Hide the evidence
   D. Post a fire fighter at the entrance to the area of fire origin

_____ 5. Pulling salvage covers on top of the contents within the area of fire origin potentially introduces a source of
   A. Cross-contamination
   B. Additional fuel
   C. Damaged salvageable goods
   D. Confusing fire patterns

_____ 6. Management of overhaul is crucial, especially within the area of
   A. Ceiling or wall damage
   B. Holes in the floor
   C. Fire origin
   D. Initial entry

_____ 7. Suppression personnel should fuel any power tools that they may be using outside of the fire scene to avoid
   A. Spilling fuel into the area under investigation
   B. Adding fuel to the fire
   C. Fueling a hot tool
   D. Vapors in the fire area

_____ 8. Fire investigator tools should be kept separate from other fire department equipment and must never be coated with
   A. Flammable solvents
   B. Rust-preventive material
   C. Paint
   D. Dust

_____ 9. The collection method of physical evidence is governed by the evidence
   A. Type
   B. Form and size
   C. Condition
   D. All of the above

_____ 10. After the physical evidence has been documented with field notes and photographs and fixed in a diagram, it is ready to be
   A. Destroyed
   B. Put back in the original position
   C. Collected and preserved
   D. Overhauled

Fill-in

Read each item carefully and then complete the statement by filling in the missing word(s).

1. The first stage of preservation of potential physical evidence on a fire ground begins with the _____ operation itself.

2. Physical evidence is something that can be observed or physically handled by a _____ _____ _____ and differs from other forms of trial evidence.

3. Although the most important physical evidence in a fire investigation is generally found within the _____ _____ _____, important evidence can also be found elsewhere within the fire scene and frequently outside the fire scene as well.

4. Inadvertent or intentional spoliation of physical evidence could potentially expose the investigator to _____ _____.

5. The investigator should think of the entire scene as _____ _____ until the process of investigation can narrow down the key areas.

6. Before any evidence is removed from the area of fire origin, it needs to be analyzed and photographically documented and _____ _____ _____ _____ that indicates its location and position before it is collected.

7. In cases in which there are two or more areas of fire origin in an incendiary fire, the investigator should always change gloves and _____ _____ _____ between sites.

8. _____ _____ _____ _____ includes, but is not limited to, finger and palm prints, bodily fluids such as blood and saliva, hair and fibers, footwear impressions, tool marks, soils and sand, woods and sawdust, glass, paint, metals, handwriting, questioned documents, and general types of trace evidence.

Chapter 14: Physical Evidence

Vocabulary
Define the following terms using the space provided.

1. Accelerant:

2. Cross-contamination:

3. Demonstrative evidence:

4. Headspace:

5. Traditional forensic physical evidence:

Short Answer
Complete this section with short written answers, using the space provided.

1. Identify eight key properties of common ignitable liquids.

2. List six desirable ignitable liquid collection areas.

_____
_____
_____
_____

3. Describe the steps recommended to recover ignitable liquid containers suspected to be evidence in a suspicious fire.

_____
_____
_____
_____
_____

4. Describe acceptable procedures for protecting and preserving a fire investigation scene.

_____
_____
_____
_____

5. List the data that should be included when marking an item of physical evidence, tag, or package for identification.

_____
_____
_____
_____
_____

Fire Alarms

The following case scenario will give you an opportunity to explore the concerns associated with fire investigation. Read the scenario and then answer each question in detail.

You are your department's fire investigator and have responded with suppression companies to a single-family residential structure fire. You arrive at the working fire at the same time the first fire companies report on scene.

1. What should be included in your initial observations and primary search for evidence?

_____
_____
_____
_____
_____

2. As far as fire cause, what should the suppression companies be taking note of and be concerned with during their suppression activities?

3. What is your role as the investigator during the suppression activities?

Additional Activity

1. On your arrival at a residential structure fire, the incident commander hands you a gasoline can. He states the steel can was found in the front room of the structure by the initial attack team company officer.

   Prepare a form that will help positively identify the evidence and maintain a chain of custody and accurate historical accounting of the steel can.

Origin Determination

Workbook Activities

The following activities have been designed to help you. Your instructor may require you to complete some or all activities as a regular part of your investigator training program. You are encouraged to complete any activity your instructor does not assign as a way to enhance your learning in the classroom.

Chapter Review

The following exercises provide an opportunity to refresh your knowledge of this chapter.

Matching

Match each of the terms in the left column to the appropriate definition in the right column.

_____ 1. Point of origin

_____ 2. Fire spread analysis

_____ 3. Total burn

_____ 4. Isochar

_____ 5. Sequential pattern analysis

_____ 6. Area of origin

_____ 7. Depth of char surveys

_____ 8. Arc mapping

_____ 9. Fire scene reconstruction

_____ 10. Initial scene assessment

A. The exact physical location, as far as the investigator can determine, where a heat source and a fuel come into contact with each other and a fire begins.

B. Provides an overall look at the structure and surrounding area.

C. Refers to the room, building, or general area in which the point of origin is located.

D. Measurements of the relative depth of char on identical fuels are plotted on a detailed scene diagram to determine locations within a structure that were exposed longest to a heat source.

E. Process of removal of debris and replacement of contents or structural elements in their prefire positions.

F. A line on a diagram connecting equal points of char depth.

G. Application of principles of fire science to the analysis of fire pattern data (including fuel packages and geometry, compartment geometry, ventilation, fire suppression operations, witness information, etc.) to determine the origin, growth, and spread of a fire.

H. A fire that has consumed nearly all of the fuel (contents and/or structure).

I. Locations of electrical arcing are identified and plotted on a diagram of the affected area of the structure. The spatial relationship of the arc sites can create a pattern and help establish the sequence of damage.

J. Must be conducted to determine whether the physical damage and other available data are consistent with the origin hypothesis.

CHAPTER 15

Multiple Choice

Read each item carefully and then select the best response.

_____ 1. Relevant information relating to origin can be obtained from
   A. Fire patterns
   B. Arc mapping
   C. Heat and flame vector analysis
   D. All of the above

_____ 2. The methodology recommended for examination of a fire scene includes a(n)
   A. Rapid scene assessment
   B. Initial scene assessment
   C. No documentation of findings
   D. Contamination of evidence

_____ 3. A fire spread analysis must be conducted to determine whether the physical damage and other available data are consistent with the
   A. Origin hypothesis
   B. Chief's decree
   C. Owners statement
   D. SWAG

_____ 4. The first order of business in the initial scene assessment is the
   A. Evidence removal
   B. Witness interview
   C. Safety assessment
   D. Scene excavation

_____ 5. Detector activation and sprinkler head response can be utilized to help identify a(n)
   A. Fire set
   B. Cause of origin
   C. Area of origin
   D. Confusing fire pattern

_____ 6. Identifying the prefire location and orientation of contents is essential to
   A. Scene photography
   B. Fire scene reconstruction
   C. Scene safety
   D. Heavy equipment use

_____ 7. The origin hypothesis should explain not only where the fire started but
   A. Who set the fire
   B. What accelerant was used
   C. Why the fire was set
   D. How it spread throughout the structure

8. Fire patterns are generated by either the spread of the heat/flames or the
   A. Intensity of burning
   B. Type of accelerant
   C. Floor coverings
   D. Size of the structure

9. The key to appropriate use of depth of char is documenting the relative depth of charring from point to point and locating places where damage was most severe due to exposure, ventilation, or
   A. Extinguishment attempts
   B. Overhaul efforts
   C. Smoke conditions
   D. Fuel placement

10. A technique in which the investigator uses the identification of locations of electrical arcing to help determine the area of origin:
    A. Char depth analysis
    B. Isochar
    C. Arc mapping
    D. Fire pattern analysis

Fill-in

Read each item carefully and then complete the statement by filling in the missing word(s).

1. The inability to determine the point of origin _____ _____ eliminate the ability to develop a credible and defensible origin and cause hypothesis.

2. Adequate _____ _____ _____ is essential in leading the investigator to the correct analysis.

3. Where heavy equipment must be operated within the scene, the equipment should initially be positioned _____ the area of interest.

4. The _____ _____ _____ phase is important because it can allow for a strong visualization of how the fire developed and spread.

5. Reliance on a single fire pattern, except in cases of extremely small fires, will likely result in _____ origin hypotheses.

6. If a fire originates in one part of a compartment and spreads to involve ignitable liquid stored elsewhere in that compartment the investigator may be misled by the _____ _____ from the burning ignitable liquid.

7. The _____ _____ _____ _____ analysis is a process used to assist with fire spread analysis and origin determination.

8. In the absence of _____ _____ to establish the origin, an eyewitness to the early stages of the fire may be able to provide reliable evidence.

Vocabulary

Define the following terms using the space provided.

1. Arc survey diagrams:

   _____

   _____

   _____

2. Area of origin:

_____
_____
_____

3. Depth of char surveys:

_____
_____
_____

4. Depth of calcination surveys:

_____
_____
_____

5. Fire spread analysis:

_____
_____
_____

6. Isochar:

_____
_____
_____

7. Point of origin:

_____
_____
_____

8. Safety assessment:

_____
_____
_____

9. Sequential pattern analysis:

_____
_____
_____

10. Total burn:

Short Answer
Complete this section with short written answers, using the space provided.

1. Identify six sources where relevant information relating to fire origin can be obtained.

2. Describe the process for data collection, as it relates to fire origin determination.

3. Identify building features that should be documented when conducting an in-depth building exterior examination.

4. Identify building features that should be documented when conducting an in-depth building interior examination.

5. Describe the process and concerns with conducting a debris removal and reconstruction investigation phase.

Fire Alarms

The following case scenario will give you an opportunity to explore the concerns associated with fire investigation. Read the scenario and then answer each question in detail.

You have been assigned to investigate a two-story commercial structure fire. The first arriving company's initial size-up included the information that fire was blowing out the front door (A side) of a two-story cinder block structure, with flames and smoke also blowing out an A side second story window. The companies made an aggressive interior attack and extinguished the fire.

1. Describe the recommended techniques for an initial scene assessment that will aid in determining the origin of this fire.

   _____
   _____
   _____
   _____
   _____

2. Explain issues of evidence preservation, contamination, and spoliation during debris removal and scene reconstruction.

   _____
   _____
   _____
   _____
   _____

Fire Cause Determination

Workbook Activities

The following activities have been designed to help you. Your instructor may require you to complete some or all activities as a regular part of your investigator training program. You are encouraged to complete any activity your instructor does not assign as a way to enhance your learning in the classroom.

Chapter Review

The following exercises provide an opportunity to refresh your knowledge of this chapter.

Matching

Match each of the terms in the left column to the appropriate definition in the right column.

_____ 1. Thermal inertia   **A.** Identification of the first fuel ignited

_____ 2. Ignition sequence   **B.** Process through which all potential ignition sources in the area of origin are identified and then considered in light of the physical properties of the first fuel ignited, fundamental scientific principles, and other available data

_____ 3. Ignition source analysis   **C.** The product of thermal conductivity, density, and specific heat (or heat capacity)

_____ 4. Fuel analysis   **D.** The sequence of events and circumstances that allow the initial fuel and the ignition source to come together and result in a fire

_____ 5. Negative corpus   **E.** Most common oxidizing agent

_____ 6. Atmospheric oxygen   **F.** Improper use of the process of elimination

Multiple Choice

Read each item carefully and then select the best response.

_____ 1. Identification of the first fuel ignited is an important part of what step?
   A. Testing the hypothesis
   B. Eliminating arson
   C. Determining physical evidence
   D. Process of elimination

CHAPTER 16

_____ 2. With like fuels, the fuel with the higher surface-to-mass ratio will
   A. Ignite slower
   B. Ignite at a higher temperature
   C. Ignite at a lower temperature
   D. Require less energy to ignite

_____ 3. Where is an ignition source when a fire starts?
   A. In the "rich" part of the flammable range
   B. In the "lean" part of the flammable range
   C. At the point of origin
   D. Away from the fuel supply

_____ 4. A hypothesis based on the absence of physical evidence is one that should be
   A. Easy to prove
   B. Approached with caution
   C. Easy to disprove
   D. Removed from consideration

_____ 5. Which type of evidence must be present for a hypothesis to be tested?
   A. Evidence of an ignition source
   B. Physical evidence
   C. Eyewitness testimony
   D. Photographic evidence

_____ 6. An ignition source must be able to generate energy sufficient to
   A. Activate an oxidizer
   B. Generate gases
   C. Raise the fuel to its ignition temperature
   D. Begin a smolder stage

_____ 7. What should be done with oxidant residue after a fire?
   A. Collected
   B. Burned off
   C. Shoveled out with the refuse
   D. Allowed to dissipate

_____ 8. What does a fire investigator establish by identifying the fire ignition sequence?
   A. Fuel used
   B. Motive
   C. Cause hypothesis
   D. Cause and responsibility

104 FIRE INVESTIGATOR: PRINCIPLES AND PRACTICE TO NFPA 921 AND 1033

_____ 9. What is the appropriate way to apply the scientific method to testing the cause hypothesis?
   A. Attempt to prove the hypothesis
   B. Attempt to disprove the hypothesis
   C. Identify cause and responsibility
   D. Identify the origin

_____ 10. Evaluation of the competence of a hypothesized ignition source requires identification of
   A. The origin
   B. The oxidant
   C. The initial fuel
   D. The extinguishing agent

Fill-in

Read each item carefully and then complete the statement by filling in the missing word(s).

1. In any investigation of a fire cause, the investigator should consider all reasonable _____ _____ and use the scientific method to test the hypothesis of origin and cause.

2. In fire scenes where the investigator has a clearly defined point of origin and there is no identifiable ignition source, the lack of evidence may afford a _____ for consideration.

3. A hypothesis based on the absence of _____ _____ is one that should be approached with caution.

4. Simply eliminating all of the known potential ignition sources in the area of origin in and of itself is _____ _____ to conclude that a fire was intentionally set.

5. When diffuse fuels are the initial fuel, the point of origin may be _____ from the location of other fuels that sustain combustion.

6. Any identified ignition source must have sufficient temperature and energy and must have the ability to raise the fuel to its _____ _____.

7. The term _____ _____ is used to describe the response of fuel to the energy that is impacting it.

8. Although the first fuel ignited may have an ignition temperature that is well within the temperatures produced by the ignition source, the _____ of the first fuel ignited is important to consider.

9. The ignition source for ignitable vapors can be more difficult to identify because of the ability of the vapor cloud to _____.

10. Hypotheses that cannot be disproved may be considered either _____ or _____.

Vocabulary

Define the following terms using the space provided.

1. Fuel analysis:

   _____
   _____
   _____

2. Ignition sequence:

_____
_____
_____

3. Thermal inertia:

_____
_____
_____

Short Answer
Complete this section with short written answers, using the space provided.

1. Outline the three-step process used to determine if an ignition source has sufficient temperature and energy to ignite the fuel at the point of origin.

_____
_____
_____
_____
_____
_____
_____

2. List the data essential to determine fire cause.

_____
_____
_____
_____
_____
_____

3. Identify six factors to be considered in developing an ignition sequence.

_____
_____
_____
_____
_____
_____

4. Define the two levels of certainty an investigator may establish, based on his or her confidence in the data collected.

___

5. List three common oxidants that may be found at a fire scene.

___

Fire Alarms

The following case scenario will give you an opportunity to explore the concerns associated with fire investigation. Read the scenario and then answer each question in detail.

You are investigating a kitchen fire and place the point of origin on the countertop under the cabinets. Your initial investigation leads you to believe no flammable liquids or accelerants were involved. You interview the occupant and discover there were several appliances on the countertop, including a mixer, a radio, a coffee maker, and a microwave oven.

1. How would you assess the potential ignition sources? Use your experience.

___

Additional Activity

1. You are investigating a bedroom fire and are trying to establish the circumstances that allowed the initial fuel and the ignition source to come together. You believe the fire was intentionally set but need to establish the ignition sequence to help verify your suspicions. Identify six factors you should consider to establish the events that occurred before and during actual initial ignition.

   _____
   _____
   _____
   _____
   _____
   _____
   _____

Analyzing the Incident for Cause and Responsibility

Workbook Activities

The following activities have been designed to help you. Your instructor may require you to complete some or all activities as a regular part of your investigator training program. You are encouraged to complete any activity your instructor does not assign as a way to enhance your learning in the classroom.

Chapter Review

The following exercises provide an opportunity to refresh your knowledge of this chapter.

Matching

Match each of the terms in the left column to the appropriate definition in the right column.

_____ 1. Accidental fires   A. Fires intentionally ignited under circumstances in which the person knows that the fire should not be ignited.

_____ 2. Natural fires   B. The accountability of a person or other entity for the event or sequence of events that caused the fire or explosion, spread of the fire, bodily injuries, loss of life, or property damage.

_____ 3. Incendiary fires   C. Fires caused without direct human intervention or action, such as fires resulting from lightning, earthquake, and wind.

_____ 4. Undetermined fires   D. A logical, systematic examination of an item, component, assembly, or structure and its place and function within a system, conducted to identify and analyze the probability, causes, and consequences of potential and real failures.

_____ 5. Responsibility   E. The cause of the fire cannot be proven to an acceptable level of certainty.

_____ 6. Failure analysis   F. Those fires that are not the result of a deliberate (intentional) act.

_____ 7. Suspicious   G. A term that denotes a crime and is therefore determined by judicial process.

_____ 8. Arson   H. An unacceptable classification for a fire cause. Refers to a level of proof or level of certainty.

Multiple Choice

Read each item carefully and then select the best response.

_____ 1. This area concerns factors related to human contribution (i.e., act or omission) to the any of the features listed above. For example, combustible materials might have been stored too close to a heat source:
   A. Cause of the fire or explosion
   B. Cause of damage to property resulting from the incident
   C. Degree to which human involvement contributed to cause
   D. Cause of bodily injury or loss of life

CHAPTER 17

_____ 2. This area identifies the elements of a cause: the heat source, the first fuel ignited, the oxidizer, and the conditions or circumstances that allowed these components to come together and result in a fire or explosion:
   A. Cause of the fire or explosion
   B. Cause of damage to property resulting from the incident
   C. Degree to which human involvement contributed to cause
   D. Cause of bodily injury or loss of life

_____ 3. This area considers the factors that are responsible for fire spread from the origin. These factors may include the combustibility of contents or construction materials, the adequacy or inadequacy of passive (e.g., fire walls and fire doors) and active (e.g., water sprinklers) fire protection systems, and structural compliance to applicable fire or building codes:
   A. Cause of the fire or explosion
   B. Cause of damage to property resulting from the incident
   C. Degree to which human involvement contributed to cause
   D. Cause of bodily injury or loss of life

_____ 4. This area identifies factors related to human injuries or deaths. The factors may include analysis of fire alarm or smoke detection systems, means of egress, or the role of materials that emit products that are harmful to humans during a fire:
   A. Cause of the fire or explosion
   B. Cause of damage to property resulting from the incident
   C. Degree to which human involvement contributed to cause
   D. Cause of bodily injury or loss of life

_____ 5. Cause classification is the culmination of identifying the four elements of the fire and then categorizing them according to the generally accepted definitions. The four elements are
   A. Fire set, cause of origin, area of origin, extent of burn
   B. Fuel, oxidizer, heat, chain reaction
   C. Intensity of burning, type of accelerant, floor coverings, size of the structure
   D. Ignition source, first fuel ignited, oxidizer, and ignition factor

_____ 6. The cause of an explosion or fire can be classified into four general categories:
   A. Extinguishment, overhaul, smoke, fuel
   B. Fuel, oxidizer, heat, chain reaction
   C. Accidental, natural, incendiary, undetermined
   D. Ignition source, first fuel ignited, oxidizer, and ignition factor

_____ 7. Those fires that are not the result of a deliberate (intentional) act:
   A. Natural
   B. Accidental
   C. Incendiary
   D. Undetermined

_____ 8. Fires that ignite without human intervention:
   A. Natural
   B. Accidental
   C. Incendiary
   D. Undetermined

_____ 9. Fires that have not yet been investigated, fires that are under investigation, and fires that have been investigated and the cause is not proven to an acceptable level of certainty:
   A. Natural
   B. Accidental
   C. Incendiary
   D. Undetermined

_____ 10. Fires that result from deliberate actions in circumstances in which the person starting the fire knows that he or she should not start a fire:
   A. Natural
   B. Accidental
   C. Incendiary
   D. Undetermined

Fill-in
Read each item carefully and then complete the statement by filling in the missing word(s).

1. The mindset or mental state (_____) of the fire-setter is a key element of the incendiary fire classification.

2. Fire investigations should be approached without _____. Therefore, all fire investigations theoretically start with an undetermined classification.

3. If the cause cannot be determined because the data are insufficient to support the hypothesis, then the cause must be classified as _____.

4. In limited circumstances, it may be possible to make a credible cause determination in the absence of the _____ evidence of the ignition source.

5. The investigator cannot use evidence of _____ to classify the cause.

6. At no time is a fire to be classified as _____.

7. Development and spread of heat and smoke are of great significance in identifying _____ _____ of bodily injury or death.

8. Determining responsibility frequently requires that the investigator conduct a _____ _____.

Vocabulary
Define the following terms using the space provided.

1. Incendiary fires:

_____

_____

_____

2. Natural fires:

3. Responsibility:

4. Undetermined fire:

Short Answer

Complete this section with short written answers, using the space provided.

1. Differentiate between incendiary fires and arson fires.

2. Identify and describe the four general categories to which the cause of an explosion or fire can be classified.

3. Explain the differences between cause of a fire and responsibility for a fire.

4. Identify 10 factors that can contribute to the development and spread of fire and smoke.

5. Identify 12 analytical tools used for the failure analysis process.

Fire Alarms

The following case scenario will give you an opportunity to explore the concerns associated with fire investigation. Read the scenario and then answer each question in detail.

You have been assigned to prepare and deliver a presentation on significant fires in the United States that have contributed to the evolution of building codes. Part of your presentation is to include information on how a proper failure analysis identified contributing factors to the fire and resulting loss of life.

1. Describe failure analysis.

   _____
   _____
   _____
   _____
   _____
   _____
   _____
   _____

2. What are some examples of circumstances to which failure analysis can be applied during an investigation?

   _____
   _____
   _____
   _____
   _____
   _____
   _____
   _____

Failure Analysis and Analytical Tools

Workbook Activities

The following activities have been designed to help you. Your instructor may require you to complete some or all activities as a regular part of your investigator training program. You are encouraged to complete any activity your instructor does not assign as a way to enhance your learning in the classroom.

Chapter Review

The following exercises provide an opportunity to refresh your knowledge of this chapter.

Matching
Match each of the terms in the left column to the appropriate definition in the right column.

_____ 1. Relative time

_____ 2. Hard time

_____ 3. Soft time

_____ 4. Estimated time

_____ 5. System analysis

_____ 6. Timeline

_____ 7. Fault tree

_____ 8. Heat transfer model

_____ 9. Benchmark event

_____ 10. Failure mode and effects analysis

A. Identifies an estimated or relative point in time.

B. An approximation based on information or calculations that may or may not be relative to other events or activities.

C. Event that is particularly valuable as a foundation for the timeline or may have significant relation to the cause, spread, detection, or extinguishment of a fire.

D. A logic diagram that can be used to analyze a fire or explosion. Also known as a decision tree.

E. Identifies a specific point in time that is directly or indirectly linked to a reliable clock or timing device of known accuracy.

F. A technique used to identify basic sources of failure within a system and to follow the consequences of these failures in a systematic fashion.

G. Allows the investigator to determine how heat was transferred from a source to a target by one or more of the common heat transfer modes: conduction, convection, or radiation.

H. An analytical approach that takes into account characteristics, behavior, and performance of a variety of elements.

I. Chronological order of events or activities that can be identified in relation to other events or activities.

J. A graphic or narrative representation of events related to the fire incident, arranged in chronological order.

CHAPTER 18

Multiple Choice

Read each item carefully and then select the best response.

_____ 1. When developing a timeline, incidents that can be related to a known exact time are generally referred to as
   A. Soft times
   B. Hard times
   C. Estimated times
   D. Relative times

_____ 2. Soft time is either estimated or relative and is generally provided by
   A. Dispatch
   B. IC
   C. Witnesses
   D. PD

_____ 3. An approximation based on information or calculations that may or may not be relative to other events or activities:
   A. Soft times
   B. Hard times
   C. Estimated times
   D. Scaled timeline

_____ 4. Can cover months or even years and may incorporate activities that occurred a long time before the fire:
   A. Macro evaluation of events
   B. Micro evaluation of events
   C. Scaled timeline
   D. Parallel times

_____ 5. Can be used to look at small or narrow segments of the macro timeline in detail:
   A. Macro evaluation of events
   B. Micro evaluation of events
   C. Scaled timeline
   D. Parallel times

_____ 6. Can be used to look at multiple events that occur simultaneously:
   A. Macro evaluation of events
   B. Micro evaluation of events
   C. Scaled timeline
   D. Parallel times

_____ 7. Shows the time of each event with the spacing between the time events scaled in a manner that illustrates the elapsed time between each event:
   A. Macro evaluation of events
   B. Micro evaluation of events
   C. Scaled timeline
   D. Parallel times

8. Places the conditions and chains of events that are necessary for a given fire or explosion to occur:
   A. Macro evaluation
   B. Fault tree diagram
   C. Scaled timeline
   D. Estimated times

9. Can be used to test hypotheses such as the competency of a given heat source to act as an ignition source in a given fire causation scenario:
   A. Micro evaluation
   B. Scaled timeline
   C. Relative times
   D. Heat transfer models

10. The analysis of a sprinkler system and its water supply should determine whether the system and water supply were matched to
   A. The hazard being protected
   B. Each other
   C. The weather
   D. The structure

Fill-in

Read each item carefully and then complete the statement by filling in the missing word(s).

1. Fires and explosions that are believed to be caused by reactions of known or suspected chemical mixtures can be investigated by a _____ _____ of the probable chemical mixtures and potential contaminants.

2. A _____ _____ can be critical in helping to point toward where the fire was able to have the most significant impact on the strength of the building.

3. The reason why a fire victim did not escape from a given fire scene is a _____ _____ for a fire investigator to answer.

4. Specialized fire dynamics routines and computer models can be used to test and evaluate an _____ _____ _____ _____ and can assist in the evaluation of physical and eyewitness evidence used as part of the hypothesis development process.

5. Using known input data, _____ _____ can be compared with witness accounts and physical evidence as a test of a developed hypothesis or account.

6. _____ _____ _____ _____ are simplified procedures that require minimal data to run a computer model and can often answer a narrowly focused question.

7. In forensic applications, _____ can be used to help a judge and/or jury to understand better important features of a fire scene and underlying scientific and engineering principles that caused a particular outcome in a given fire.

8. A meaningful analysis of a fire requires understanding of the _____ _____ _____, the fire growth rate, and total heat released.

Vocabulary

Define the following terms using the space provided.

1. Benchmark event:

2. Estimated time:

3. Failure mode and effects analysis (FMEA):

4. Fault tree:

5. Hard time:

6. Heat transfer model:

Short Answer

Complete this section with short written answers, using the space provided.

1. List at least four items an investigator should examine to obtain data relating to fire victims and escape from a structure.

2. Identify six fire-related factors or events computer-based fire dynamics models may be able to predict and may be useful to the fire investigator.

3. Identify four variables that can influence fire modeling results. How can these variables alter the modeling results?

4. List eight specific issues specialized fire dynamics routines results can be used to evaluate.

5. Describe the differences between zone models and CFD models. How are these types of modeling useful in a fire investigation?

Fire Alarms

The following case scenario will give you an opportunity to explore the concerns associated with fire investigation. Read the scenario and then answer each question in detail.

You are investigating the cause of a mercantile structure fire and are certain the case will end up in court. As you are collecting evidence and recording your observations you are very aware of the need for the accuracy and completeness of your evidence. Much of your information will be sent to your local arson investigation lab for verification of your hypothesis.

1. What information is required to create a timeline showing elapsed time between events related to the fire?

   _____
   _____
   _____
   _____
   _____

2. Identify three types of information required for valid modeling and testing. This is the information you should obtain and include in the data you will send to the lab.

   _____
   _____
   _____
   _____
   _____

Explosions

Workbook Activities

The following activities have been designed to help you. Your instructor may require you to complete some or all activities as a regular part of your investigator training program. You are encouraged to complete any activity your instructor does not assign as a way to enhance your learning in the classroom.

Chapter Review

The following exercises provide an opportunity to refresh your knowledge of this chapter.

Matching

Match each of the terms in the left column to the appropriate definition in the right column.

_____ 1. BLEVE
_____ 2. Chemical explosion
_____ 3. Explosive
_____ 4. Deflagration
_____ 5. Explosion
_____ 6. Low-order damage
_____ 7. Detonation
_____ 8. High-order damage
_____ 9. Mechanical explosion
_____ 10. Combustion explosion

A. A reaction that propagates at a subsonic velocity, several feet per second, and can be successfully vented.

B. A slow rate of pressure rise or low-force explosion characterized by a pushing or dislodging effect on the confining structure or container and by short missile distances.

C. Boiling liquid expanding vapor explosion.

D. A reaction that propagates at supersonic velocities, greater than 110 feet per second, and cannot be vented because of its speed.

E. That in which a chemical reaction is the source of the high-pressure fuel gas. The fundamental nature of the fuel is changed.

F. A rapid pressure rise or high-force explosion characterized by a shattering effect on the confining structure or container and long missile distances.

G. That of high-pressure gas producing a physical reaction such as the rupture of a container. The fundamental nature of the fuel is not changed.

H. That caused by the burning of combustible hydrocarbon fuels and characterized by the presence of a fuel with air as an oxidizer.

I. Any chemical compound, mixture, or device that functions by explosion.

J. The sudden conversion of potential energy (chemical or mechanical) into kinetic energy with the production and release of gases under pressure or the release of gas under pressure.

CHAPTER 19

Multiple Choice

Read each item carefully and then select the best response.

_____ 1. What type of energy is possessed by a system or object as a result of its motion?
   A. Potential
   B. Kinetic
   C. Hidden
   D. Chemical

_____ 2. What type of fire spreads by means of a flame front rapidly through a diffuse fuel such as dust, gas, or ignitable liquid vapor and without production of damaging pressure?
   A. Backdraft
   B. Flameover
   C. Flash fire
   D. Explosion

_____ 3. What are the two major types of explosions?
   A. Dust and vapor
   B. Mechanical and chemical
   C. Major and minor
   D. Gas and liquid

_____ 4. What types of explosions are caused when high-energy arcing generates sufficient heat to cause an explosion?
   A. Electrical explosions
   B. Chemical explosions
   C. Vapor explosions
   D. Dust explosions

_____ 5. Which pressure phase is more powerful and is responsible for most of the pressure damage?
   A. Negative pressure phase
   B. Positive pressure phase
   C. Hot pressure phase
   D. High-pressure phase

_____ 6. What explosion type releases heat energy and can cause secondary fires?
   A. Combustion
   B. Chemical
   C. Detonating
   D. Deflagration

_____ 7. What explosion type produces lower temperatures of longer duration?
   A. Combustion
   B. Chemical
   C. Detonating
   D. Deflagration

_____ 8. What explosion type produces extremely high temperatures of short duration?
   A. Combustion
   B. Chemical
   C. Detonating
   D. Deflagration

_____ 9. Which type of fuel generates seated explosions?
   A. Fuel gases or vapors
   B. Dust
   C. Smoke
   D. Explosives

_____ 10. What are the most commonly encountered explosions?
   A. Fuel gases or vapors
   B. Dust
   C. Smoke
   D. Explosives

_____ 11. What is the rate of flame propagation relative to the velocity of the unburned gas ahead of it?
   A. Flame speed
   B. Burning velocity
   C. Transitional velocity
   D. Detonation speed

_____ 12. What is the ratio of the average molecular weight of a given volume of gas or vapor to the same of air at the same temperature and pressure?
   A. Gas mass
   B. Specific mass
   C. Vapor density
   D. Flammable density

_____ 13. What does increasing the moisture content of dust particles do to the dust's ignition potential?
   A. Increases the minimum energy required for ignition
   B. Decreases the minimum energy required for ignition
   C. Does not affect the minimum energy required for ignition
   D. Causes the dust to burn hotter

_____ 14. High explosives are designed to produce what type effect?
   A. Burning and scorching
   B. Subsonic blast pressure
   C. Pushing and heaving
   D. Shattering

_____ 15. Where should the outer perimeter of a blast zone be established?
   A. Half the distance of the farthest piece of debris
   B. One and one-half the distance of the farthest piece of debris
   C. 500 feet from the epicenter
   D. 750 feet from the epicenter

Fill-in

Read each item carefully and then complete the statement by filling in the missing word(s).

1. The most common _____ _____ are those caused by the burning of combustible hydrocarbon fuels.

2. A _____ is a reaction that travels through the air (propagates) at subsonic velocities, several feet per second.

3. A significant difference between deflagration and detonation is _____.

4. The characteristics of _____-_____ _____ include walls bulged out or laid down, roofs lifted slightly, windows dislodged with the glass intact, and thrown-out debris that are generally large and found within a short distance from the structure.

5. The transmission of tremors through the ground is known as _____ _____.

6. When a blast pressure front hits an object, the pressure may reflect off the object, thus _____ the blast pressure front and overpressure.

7. The crater or area of greatest damage may be characterized as the _____ _____ _____ _____.

8. The size, shape, construction, volume, materials, and design of the _____ _____ greatly affect the nature of the explosion damage.

9. Lighter-than-air and heavier-than-air fuel gases that have escaped from underground piping systems _____ _____ to enter structures, resulting in fires or explosions.

10. Violent explosions can be fueled by _____ that is dispersed within the air.

Vocabulary

Define the following terms using the space provided.

1. Burning velocity:

2. Deflagration:

3. Explosion dynamics analysis:

4. Firebrands:

5. High-order damage:

6. Mechanical explosion:

7. Seismic effect:

8. Transitional velocity:

Short Answer
Complete this section with short written answers, using the space provided.

1. List the four major effects of an explosion.

2. Identify some of the forces that can shape a blast front.

3. List seven factors that control the effects of an explosion.

4. Explosion damage to structures is related to what factors?

5. What should an investigator do if he or she determines an explosion was fueled by an explosive device?

Fire Alarms

The following case scenario will give you an opportunity to explore the concerns associated with fire investigation. Read the scenario and then answer each question in detail.

You are dispatched to investigate a residential structure explosion. Upon arrival you notice debris scattered in a large area around what 30 minutes ago was a two-story house. There are areas of smoldering visible in the debris field, and companies are just beginning to hit the hot spots with hose lines.

1. What tasks would you include in your initial scene size-up?

2. What would aid you in determining the fuel source?

Additional Activity

1. Your community experienced an act of domestic terrorism. An individual had used an explosive device to disrupt a public function. Fortunately, there was little damage and no injuries. A suspect was apprehended quickly, and justice was served. The lack of damage and injury was due to the fact the terrorist used a small amount of low explosives in a paper bag.

    The chief of your department has asked you to put together a presentation for the fire department explaining the differences between low and high explosives, as well as identifying common examples of both.

    Use the information from Chapter 19 as a resource for this presentation.

Incendiary Fires

Workbook Activities

The following activities have been designed to help you. Your instructor may require you to complete some or all activities as a regular part of your investigator training program. You are encouraged to complete any activity your instructor does not assign as a way to enhance your learning in the classroom.

Chapter Review

The following exercises provide an opportunity to refresh your knowledge of this chapter.

Matching

Match each of the terms in the left column to the appropriate definition in the right column.

_____ 1. Sabotage
_____ 2. Serial arson
_____ 3. Modus operandi (MO)
_____ 4. Trailer
_____ 5. Intent
_____ 6. Spree arson
_____ 7. Motive
_____ 8. Victimology
_____ 9. Vandalism
_____ 10. Incendiary fire

A. The method to spread fire to other areas by deliberately linking them together with combustible fuels or ignitable liquids

B. An inner drive or impulse that is the cause, reason, or incentive that induces or prompts a specific behavior

C. A thorough understanding of the offender activity with the victim (or targeted property)

D. Intentional damage or destruction

E. A fire that is intentionally ignited under circumstances in which the person knows the fire should not be ignited

F. An offender who sets three or more fires with a cooling-off period between fires

G. Mischievous or malicious fire setting that results in damage to property

H. The setting of three or more fires at separate locations with no emotional cooling-off period between fires

I. Necessary to show proof of a crime and refers to the state of mind that exists at the time a person acts or fails to act

J. The method of operation used by the offender

Multiple Choice

Read each item carefully and then select the best response.

_____ 1. The setting of three or more fires at the same site or location during a limited period of time is called
   A. Serial arson
   B. Spree arson
   C. Mass arson
   D. Vandalism

_____ 2. What is the unusual behavior by an offender, beyond that necessary to commit the crime?
   A. Modus operandi
   B. Staging
   C. Intent
   D. Personation

_____ 3. Which of the following is **not** a potential indicator that somebody had prior knowledge of the fire?
   A. Blocked or obstructed entry
   B. Sabotage
   C. Presence of electrical equipment in area of origin
   D. Fires near service equipment and appliances

_____ 4. An offender who sets three or more fires with a cooling-off period between fires is called a
   A. Serial arsonist
   B. Spree arsonist
   C. Mass arsonist
   D. Vandal

_____ 5. What is the purposeful alteration of the crime scene before the arrival of police?
   A. Modus operandi
   B. Staging
   C. Intent
   D. Personation

_____ 6. The setting of three or more fires at separate locations with no emotional cooling-off period between fires is called
   A. Serial arson
   B. Spree arson
   C. Mass arson
   D. Vandalism

_____ 7. Ignitable liquids are commonly referred to as
   A. Accelerants
   B. Extinguishing agents
   C. Oxidizers
   D. Suppressants

_____ 8. Which of the following is **not** a common indicator of arson?
   A. Financial stress
   B. Multiple driving violations
   C. Overinsured properties
   D. Fires at additional properties owned by a single individual or group

_____ 9. Which is **not** an example of an incendiary device?
   A. Candles
   B. Wiring systems
   C. Molotov cocktails
   D. Water softener

_____ 10. Examples of "timed opportunity" include all the following except
   A. Natural conditions
   B. Civil unrest
   C. Faulty wiring
   D. Fire department unavailability

Fill-in
Read each item carefully and then complete the statement by filling in the missing word(s).

1. An _____ _____ is a fire that has been deliberately ignited under circumstances in which the person knows that the fire should not be ignited.

2. _____ _____ are fires with no obvious connection among them that would have allowed one fire to ignite the fuel in another area.

3. At full-room involvement, radiant heat can be expected to create burn patterns on floors that can be misinterpreted as _____ _____ patterns.

4. Closets, crawl spaces, and attics are typical areas, rooms, and spaces in which a limited number of _____ are present.

5. All known burn injuries to persons should be analyzed to determine their relationship to _____ _____ and the investigative hypothesis.

6. No conclusions should be made regarding the cause determination based on the _____ of the origin.

7. To make a fire appear to be _____, a fire-setter might set a fire near appliances.

8. To allow the fire more time to grow, the fire-setter might _____ _____ that will hinder or slow the firefighting operations.

9. If a fire suppression system or fire detection system failed, the investigator should inspect the systems for any signs of _____.

10. _____ as a motive comes into play to further a political, social, or religious cause.

Vocabulary
Define the following terms using the space provided.

1. High-temperature accelerant (HTA):

_____
_____
_____

2. Incendiary device:

3. Personation:

4. Spree arson:

5. Staging:

6. Trailer:

7. Vandalism:

8. Victimology:

Short Answer

Complete this section with short written answers, using the space provided.

1. List 10 "natural" means of fire spread.

2. Identify four household items commonly used as "trailers."

3. Identify three indicators of high-temperature accelerants.

4. List six examples of an incendiary device.

5. List the six motive classifications determined to be the most effective in identifying offender characteristics for firesetting behavior.

Fire Alarms

The following case scenario will give you an opportunity to explore the concerns associated with fire investigation. Read the scenario and then answer each question in detail.

During your initial investigation of a barn fire, you discover what appears to be an incendiary delay device. The device has not been activated and was discovered at a location remote from the fire point of origin.

1. How should you secure and document this device?

2. What would account for the location of the delay device, and how does this discovery help or hinder your investigation?

Fire and Explosion Deaths and Injuries

Workbook Activities

The following activities have been designed to help you. Your instructor may require you to complete some or all activities as a regular part of your investigator training program. You are encouraged to complete any activity your instructor does not assign as a way to enhance your learning in the classroom.

Chapter Review

The following exercises provide an opportunity to refresh your knowledge of this chapter.

Matching

Match each of the terms in the left column to the appropriate definition in the right column.

_____ 1. Lividity
_____ 2. Hypoxia
_____ 3. Photography
_____ 4. Pugilistic attitude
_____ 5. Carboxyhemoglobin (COHb)
_____ 6. Forensic anthropologist
_____ 7. Gas and liquid chromatography
_____ 8. Rule of nines

A. A crouching stance with flexed arms, legs, and fingers
B. May be used to assist in skeletal identification
C. Occurs after death and is the pooling of the blood in the lower elevations of the body caused by the effects of gravity
D. Techniques used to determine alcohol and drug levels in a victim
E. Used to estimate burn damage to victims
F. One of the most effective and quickest ways to document a scene
G. The carbon monoxide saturation in the blood
H. Condition caused by a victim breathing in a reduced oxygen environment

Multiple Choice

Read each item carefully and then select the best response.

_____ 1. A body found at a fire or explosion must be treated as
   A. Debris
   B. Evidence
   C. A viable patient
   D. Handling of a body is unimportant

_____ 2. How should possible fire patterns and blast effects on a body be documented?
   A. Photography
   B. Sketch
   C. Diagram
   D. Video

CHAPTER 21

_____ 3. Dental records and dental evidence can be compared and verified by whom?
   A. Forensic anthropologist
   B. Forensic archeologist
   C. Forensic odontologist
   D. Forensic pathologist

_____ 4. When should the area and pathway to the body be examined and assessed?
   A. During suppression activities
   B. After removal of the body
   C. Before a coroner arrives
   D. Before the removal of the body

_____ 5. Skeletal identification may need to be performed by whom?
   A. Forensic anthropologist
   B. Forensic archeologist
   C. Forensic odontologist
   D. Forensic pathologist

_____ 6. X-rays may be useful in identification of what type of evidence?
   A. CO in the system
   B. Illegal drugs in the system
   C. Foreign matter in the body
   D. Degree of burn injuries

_____ 7. How is carbon monoxide absorbed into the blood and tissue?
   A. Breathing
   B. Injection
   C. Ingestion
   D. Skin contact

_____ 8. Lividity becomes fixed in the body how many hours after death?
   A. 1 to 3
   B. 6 to 9
   C. 24 to 48
   D. 72 or more

_____ 9. What is the actual event or injury that brings about the cessation of life?
   A. Postmortem
   B. Mechanism of death
   C. Manner of death
   D. Cause of death

_____ 10. The course of events that led up to the accidental, homicidal, suicidal, natural, or undetermined death is called
   A. Postmortem
   B. Mechanism of death
   C. Manner of death
   D. Cause of death

Fill-in
Read each item carefully and then complete the statement by filling in the missing word(s).

1. The _____ of CO in the body is such that it can be tested for hours after the death.

2. Studies have shown that about _____ to _____ _____ of fire victims die from carbon monoxide poisoning.

3. _____ is caused when the temperature of the body is greatly elevated.

4. Inhalation of hot gases and various toxic gases causes _____ and inflammation of the airway.

5. _____ is caused by a victim breathing in a reduced oxygen environment.

6. Serious injury fires and explosions should be investigated as if they were _____ _____ to ensure completeness.

7. Because _____ _____ brings the skin in direct contact with the source, it is more dangerous than heat transferred by radiation or convection.

8. The injuries related to an explosion scene are classified into four groups based on the explosion effect that caused them: _____ _____, _____, _____, and _____.

Vocabulary
Define the following terms using the space provided.

1. Carboxyhemoglobin (COHb):

2. Hypoxia:

3. Lividity:

4. Pugilistic attitude:

Short Answer

Complete this section with short written answers, using the space provided.

1. List three items an investigator should include in a fire death diagram.

2. Identify six potential members of a fire death investigation team.

3. Describe the four degrees of burns.

4. Explain how the location and distribution of explosion injuries aid an investigation.

5. Identify 10 postmortem tests and documentations that provide an investigator with valuable information to aid in identifying a victim and establishing the cause and manner of death.

136 Fire Investigator: Principles and Practice to NFPA 921 and 1033

Labeling

Label the following diagrams with the correct percentage of possible burned surface area.

1. The rule of nines.

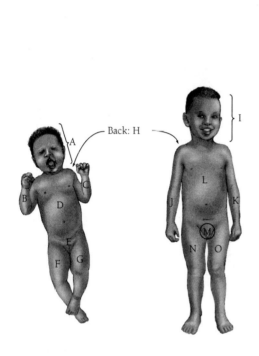
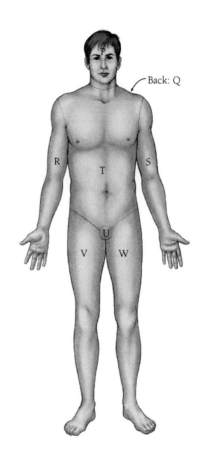

A. _____
B. _____
C. _____
D. _____
E. _____
F. _____
G. _____
H. _____
I. _____
J. _____
K. _____
L. _____

M. _____
N. _____
O. _____
P. _____
Q. _____
R. _____
S. _____
T. _____
U. _____
V. _____
W. _____

Fire Alarms

The following case scenario will give you an opportunity to explore the concerns associated with fire investigation. Read the scenario and then answer each question in detail.

You have been called to investigate an explosion and fire that occurred in a single-story deli/convenience store. The explosion killed the store owner and severely injured his partner.

1. Who would you identify as additional personnel that may be required to assist you in the investigation?

   _____
   _____
   _____
   _____
   _____
   _____

2. Outline the methods you would use to document the scene during the investigation.

   _____
   _____
   _____
   _____
   _____
   _____

Appliances

Workbook Activities

The following activities have been designed to help you. Your instructor may require you to complete some or all activities as a regular part of your investigator training program. You are encouraged to complete any activity your instructor does not assign as a way to enhance your learning in the classroom.

Chapter Review

The following exercises provide an opportunity to refresh your knowledge of this chapter.

Matching
Match each of the terms in the left column to the appropriate definition in the right column.

_____ 1. Step-down transformer
_____ 2. Interrupt current
_____ 3. Circuit breaker
_____ 4. Exemplar
_____ 5. Arc tube
_____ 6. Convection heater
_____ 7. Switch
_____ 8. Radiation heater
_____ 9. Transformer
_____ 10. Housing

A. Device used to turn the appliance on or off or to change operating conditions
B. Device that reduces AC voltage, usually 120 or 240 V AC, to a lower voltage and can be used to isolate the appliance from its power source
C. The outer shell of the appliance that contains the working components
D. Reduces the 120 V provided at the receptacle to the required voltage
E. A heater that uses a fan to move air across a hot surface, heating the air and dispersing it throughout the room
F. The value of current that the breaker must be able to carry and still operate
G. A heater that transfers heat by radiation only
H. A cylinder of fused silica/quartz with electrodes at either end that is used in most mercury and metal halide lamps
I. An exact duplicate of the appliance in question
J. Overcurrent protection device used in appliances and electrical service panels

Multiple Choice
Read each item carefully and then select the best response.

_____ 1. What appliance uses a magnetron to generate radio wave energy?
   A. Radiation heater
   B. Electric blanket
   C. Microwave oven
   D. Space heater

CHAPTER 22

_____ **2.** After an origin area has been identified and one or more appliances have been found in the origin area, the appliance should be
   A. Documented
   B. Moved
   C. Overhauled
   D. Destroyed

_____ **3.** The investigator should gather together the various appliance components that may have been moved during the fire or firefighting operations and
   A. Clean them with solvents
   B. Plug them in
   C. Give them to the owner
   D. Reconstruct the appliance

_____ **4.** If an appliance shows more severe damage than surrounding items, this could indicate that the fire
   A. Originated at the appliance
   B. Was set by an arsonist
   C. Began away from the appliance
   D. Involved flammable liquids

_____ **5.** How could a dishwasher most likely contribute to a fire?
   A. Ignition of a nearby combustible material
   B. Caused by water/moisture or lack of it
   C. A short circuit at the power cord
   D. Circuit board connection problems

_____ **6.** How could a freezer most likely contribute to a fire?
   A. Ignition of a nearby combustible material
   B. Caused by water/moisture or lack of it
   C. A power cord crimped or frayed by a recent move
   D. Circuit board connection problem

_____ **7.** What conditions must be met for determination of whether an electrical fault or overload occurred and whether sufficient heat was generated to cause ignition of the known surrounding materials?
   A. The lines to the appliance were energized
   B. The electricity was connected
   C. The appliance was turned on
   D. All of the above

_____ **8.** Because of its strength and durability, most metal appliance housings are fabricated from
   A. Aluminum
   B. Steel or stainless steel
   C. Zinc
   D. Brass

____ 9. Plastic housings, or housings made from carbon and other elements, are used in appliances that do not normally operate at
   A. High speeds
   B. Low temperatures
   C. High temperatures
   D. Low speeds

____ 10. Loose-fitting electrical plugs can create a(n)
   A. Resistive heating connection
   B. Oxidation link
   C. Mechanical heating connection
   D. Fault protector

Fill-in
Read each item carefully and then complete the statement by filling in the missing word(s).

1. The remains of batteries are usually found after the fire, and it is important to determine what they _____ _____ _____.

2. Compared with lead-acid, the lithium-metal-polymer battery has over three times the _____ _____.

3. Post-fire examination of a _____ can sometimes determine its state and thus whether the appliance was energized at the time of the fire.

4. Appliances such as a portable electric heater typically use a _____ switch to ensure that the appliance is operated in its correct designed position.

5. Appliances with heating elements are designed to maintain a distance between the element and _____ _____.

6. Lighting types normally found are incandescent and _____.

7. Any lights installed in bathrooms are susceptible to moisture and dust accumulating at the terminals, which can lead to _____ _____.

8. Common failures of magnetic ballasts that initiate fires include _____ penetrations into combustible ceiling materials.

Vocabulary
Define the following terms using the space provided.

1. Arc tube:

2. Exemplar:

3. Radiation heater:

4. Step-down transformer:

5. Transformer:

6. Triac:

Short Answer
Complete this section with short written answers, using the space provided.

1. What properties do the following sensors sense, and what type of appliances do they typically control?
   A. Thermocouple
   B. Flame sensor
   C. Current sensor

2. Identify four potential causes of dishwasher fires.

3. Identify three conditions known to cause fire in fan-forced space heaters.

_____
_____
_____

4. Identify three conditions known to cause fire in a clothes dryer.

_____
_____
_____

5. Identify three indicators that an appliance caused a fire.

_____
_____
_____

Labeling

Label the following diagrams with the correct terms.

1. Refrigerator components.

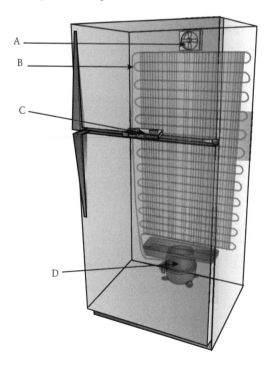

A. _____
B. _____
C. _____
D. _____

2. Clothes dryer components.

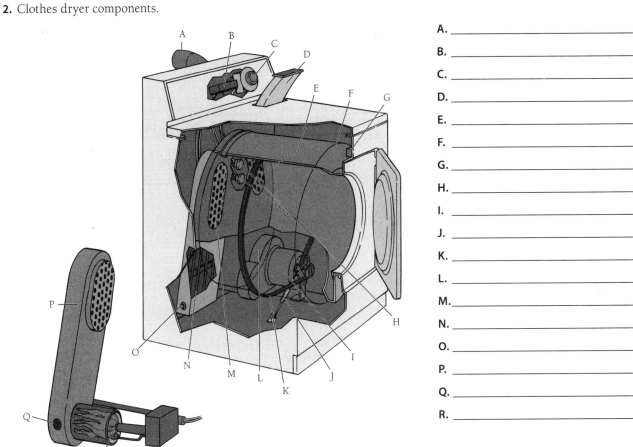

A. _____
B. _____
C. _____
D. _____
E. _____
F. _____
G. _____
H. _____
I. _____
J. _____
K. _____
L. _____
M. _____
N. _____
O. _____
P. _____
Q. _____
R. _____

Fire Alarms

The following case scenarios will give you an opportunity to explore the concerns associated with fire investigation. Read each scenario and then answer each question in detail.

1. You are investigating a fire in a residential apartment. The room of origin appears to be a laundry room. What indicators would you look for to determine if the washing machine was the cause of this fire?

_____
_____
_____
_____
_____

2. You are investigating a fire in a single-story mercantile building. The fire appears to have begun in a suspended ceiling light fixture. What should you examine regarding the light fixture?

_____
_____
_____
_____
_____

Motor Vehicle Fires

Workbook Activities

The following activities have been designed to help you. Your instructor may require you to complete some or all activities as a regular part of your investigator training program. You are encouraged to complete any activity your instructor does not assign as a way to enhance your learning in the classroom.

Chapter Review

The following exercises provide an opportunity to refresh your knowledge of this chapter.

Matching

Match each of the terms in the left column to the appropriate definition in the right column.

_____ 1. Electronic (or engine) control module (ECM)
_____ 2. Hybrid vehicle
_____ 3. CNG
_____ 4. Event data recorder (EDR)
_____ 5. Catalytic converter
_____ 6. LNG
_____ 7. Fuel rail
_____ 8. Total burn
_____ 9. Airbags
_____ 10. Turbocharger

A. Compressed natural gas
B. A device in the exhaust system that exposes exhaust gases to a catalyst metal to promote oxidation of hydrocarbon materials in the exhaust gas
C. A device to record data before and after a crash event
D. An internal passage or external tube connecting a pressurized fuel line to individual fuel injectors
E. A vehicle that uses a combination of internal combustion and electric motors for propulsion
F. Fire that has consumed all, or nearly all, combustible materials
G. Liquefied natural gas
H. An electronic device that controls engine operation parameters, including fuel delivery, throttle control, and safety systems operations
I. An exhaust-driven device that compresses intake air to increase engine power
J. Supplemental restraint systems

Multiple Choice

Read each item carefully and then select the best response.

_____ 1. Gasoline, when blended with what chemical, will change the properties of the blended fuel slightly, raising the autoignition temperature and slightly lowering the hot surface ignition temperature?
   A. Ethanol
   B. Methanol
   C. Hydrogen
   D. Propane

CHAPTER 23

_____ 2. The vacuum/low-pressure carbureted systems are most often seen on what type of vehicles?
   A. Hybrid vehicles
   B. Diesel trucks
   C. Older automobiles
   D. Fuel-injected systems

_____ 3. The high-pressure fuel-injected systems are most often seen on what type of vehicles?
   A. Lawn tractors
   B. Pick-up trucks
   C. Older automobiles
   D. Newer modern automobiles

_____ 4. Mechanical sparks capable of igniting fuel has been noted from what type of contact?
   A. Aluminum to pavement
   B. Metal to pavement
   C. Tire to pavement
   D. Glass to pavement

_____ 5. Turbochargers can be found on what type of engines?
   A. Gasoline
   B. Diesel
   C. CNG
   D. Both A and B are correct

_____ 6. What are the hottest points on the surface of the engine?
   A. Turbochargers and the exhaust manifolds
   B. Alternator
   C. Fuel rail
   D. Carburetor

_____ 7. During normal operation of a properly maintained and operating vehicle, how hot can the temperatures that enter the catalytic converter measure?
   A. 212°F
   B. 350°F
   C. 650°F
   D. 1150°F

_____ 8. Most of the interior seats and padding provide what kind of fuel load in a vehicle fire?
   A. No fuel load
   B. Significant fuel load
   C. Fuel load is undetermined
   D. Depends on ignition source

_____ 9. The first step in the investigation of a vehicle fire is to determine
   A. Whether the vehicle has been stolen
   B. Vehicle ownership
   C. A fire cause
   D. An area of fire origin

_____ 10. High-voltage circuits and harnesses on hybrid vehicles are what color for identification?
   A. Orange
   B. Safety yellow
   C. Red
   D. Electric blue

Fill-in

Read each item carefully and then complete the statement by filling in the missing word(s).

1. Gaseous fuels for motor vehicles are commonly _____ or _____.

2. Gasoline is often blended with _____ for use as motor fuel.

3. The presence of melted metals in a vehicle is not necessarily indicative of the presence of an _____ _____ accelerant.

4. Most metals need to be in powder or melted form to burn; however, _____ _____, which is present in some vehicles, burns vigorously once it is ignited by a competent external heat source.

5. In most cases, the sources of ignition energy in vehicles are the same as those associated with _____ _____.

6. The primary source of electrical energy in a vehicle that is _____ _____ is the battery.

7. Damaged insulation can result in _____ _____ when a charged wire comes into contact with a grounded surface.

8. Engine oil, power steering fluid, and brake fluid can ignite when in contact with heated _____ _____ components.

9. RVs often have _____ _____ or _____ _____ systems for heating, refrigeration, and cooking.

10. A more complete examination of the vehicle can often be _____ _____ after removal from the fire scene.

Vocabulary

Define the following terms using the space provided.

1. Catalytic converter:

_____
_____
_____

2. Electronic (or engine) control module (ECM):

_____
_____
_____

3. Event data recorder (EDR):

_____
_____
_____

Short Answer
Complete this section with short written answers, using the space provided.

1. List nine factors that affect ignition of liquids by a hot surface.

_____
_____
_____
_____
_____
_____

2. List eight informational items that may be helpful to an investigator during a vehicle fire investigation.

_____
_____
_____
_____
_____
_____

3. List nine operating systems and equipment that are unique to RVs.

_____
_____
_____
_____
_____
_____

4. Identify five fuel sources common to mass transit vehicles.

_____
_____
_____
_____
_____

5. Describe the two different types of electrical systems common to hybrid vehicles.

_____
_____
_____
_____
_____
_____

Fire Alarms

The following case scenario will give you an opportunity to explore the concerns associated with fire investigation. Read the scenario and then answer each question in detail.

An engine company has requested you to respond to a suspicious vehicle fire. The officer believes the automobile may have been stolen, and a fire was set to cover the crime.

1. What are the potential ignition sources for an automobile fire?

   _____
   _____
   _____
   _____
   _____

2. How would you document the scene before the vehicle is towed?

   _____
   _____
   _____
   _____
   _____

Wildfire Investigations

Workbook Activities

The following activities have been designed to help you. Your instructor may require you to complete some or all activities as a regular part of your investigator training program. You are encouraged to complete any activity your instructor does not assign as a way to enhance your learning in the classroom.

Chapter Review

The following exercises provide an opportunity to refresh your knowledge of this chapter.

Matching

Match each of the terms in the left column to the appropriate definition in the right column.

____ 1. Understory vegetation

____ 2. Aspect

____ 3. Fulgurites

____ 4. Duff

____ 5. Hot sets

____ 6. Prescribed fire

____ 7. Coniferous litter

____ 8. Crown

____ 9. Spot fires

____ 10. Slope

A. The layer of decomposing organic materials lying below the litter layer of freshly fallen twigs, needles, and leaves and immediately above the mineral soil.

B. Fire started in which the device is removed by the individual starting the fire.

C. Twigs, needles, or leaves of a tree or bush.

D. The area under a forest or brush canopy that grows at the lowest height level. Plants in the understory consist of a mixture of seedlings, saplings, shrubs, grasses, and herbs.

E. The direction the slope faces (N, E, S, W).

F. Spot fires are projections of flaming or burning particles (firebrands) that are found ahead of the flame front.

G. A slender, usually tubular, body of glassy rock produced by electrical current striking and then fusing dry sandy soil.

H. The steepness of land or a geographical feature that can greatly influence fire behavior.

I. A fire that resulted from intentional ignition by a person or a naturally caused fire that is allowed to continue to burn according to approved plans to achieve resource management objectives.

J. Primarily needles dropped from coniferous trees but may also include branches, bark, and cones.

CHAPTER 24

Multiple Choice
Read each item carefully and then select the best response.

_____ 1. What are all green or dead materials located in the forest canopy?
   A. Ground fuels
   B. Surface fuels
   C. Aerial fuels
   D. Suspended fuels

_____ 2. What are all combustible materials below the surface litter, including duff, tree or shrub roots, punky wood, peat, and sawdust, that normally support a glowing combustion without flame?
   A. Ground fuels
   B. Surface fuels
   C. Aerial fuels
   D. Suspended fuels

_____ 3. What are all flammable materials just above the ground?
   A. Ground fuels
   B. Surface fuels
   C. Aerial fuels
   D. Suspended fuels

_____ 4. Which is caused by pressure differences in the atmosphere that create weather patterns?
   A. Fire winds
   B. Meteorological winds
   C. Diurnal winds
   D. Foehn winds

_____ 5. What winds occur from the effects of the air being heated by the sun during the day and then cooled during the night?
   A. Fire winds
   B. Meteorological winds
   C. Diurnal winds
   D. Foehn winds

_____ 6. These winds are created when air is pulled into the bottom of a powerful convection column to replace the rising warmer air:
   A. Fire winds
   B. Meteorological winds
   C. Diurnal winds
   D. Foehn winds

_____ 7. These are the result of air flowing between pressure gradients:
   A. Fire winds
   B. Meteorological winds
   C. Diurnal winds
   D. Foehn winds

8. What is the portion of a fire that is moving most rapidly, subject to influences of slope and other topographic features?
   A. Fire flanks
   B. Fire heel
   C. Fire winds
   D. Fire head

9. This is located at the opposite side of the fire from the head; this part of the fire is less intense and is easier to control:
   A. Fire flanks
   B. Fire heel
   C. Fire winds
   D. Fire head

10. The parts of a fire's perimeter that are roughly parallel to the main direction of spread are the
    A. Fire flanks
    B. Firing out
    C. Fire break
    D. Fire storm

11. Any natural or human-made barrier used to stop the spread or reroute the direction of the fire by separating the fuel from the fire is a
    A. Fire wind
    B. Fire out
    C. Fire break
    D. Fire storm

12. The process of burning the fuel between a fire break and the approaching fire to extend the width of the fire barrier is called
    A. Fire wind
    B. Firing out
    C. Fire break
    D. Fire storm

13. A natural phenomenon that attains such intensity it creates and sustains its own wind system:
    A. Fire wind
    B. Fire out
    C. Fire break
    D. Fire storm

14. Wildfires deliberately and/or maliciously set with the intent to damage or defraud are
    A. Prescribed fire
    B. Accidental fire
    C. Incendiary fire
    D. Spot fires

15. What is a fire for which the proven cause does not involve an intentional human act to ignite or spread fire into an area where the fire should not be?
    A. Prescribed fire
    B. Accidental fire
    C. Incendiary fire
    D. Spot fires

Chapter 24: Wildfire Investigations

Fill-in
Read each item carefully and then complete the statement by filling in the missing word(s).

1. Fuels are classified as _____ _____, which include all flammable materials lying on or in the ground, _____ _____, which include all flammable materials just above the ground, and _____ _____, which include all green and dead materials located in the upper forest canopy.

2. Grass, weeds, and other small plants are ground fuels that influence the rate of fire spread, based primarily on their degree of _____.

3. Radiant heat is the dominant heat transfer method in _____ _____ _____, composed of brush and grass on level surfaces.

4. Cooler air temperatures result in _____ fuel moistures and _____ fire spread.

5. Fire crews should attempt to protect the potential _____ _____ _____ so that the fire's origin and cause can be accurately determined.

6. A fire burning uphill or with the wind creates a _____ _____ that slopes to a greater degree than the ground slope.

7. The _____ _____ is one of the best techniques for inspecting a wildfire area in great detail.

8. The action of a lightning strike causes _____ by melting sand in the soil near the base of the tree or area of contact with the ground.

9. An _____ _____ is one that is deliberately ignited by a person who knows that the fire should not be ignited.

10. A _____ _____ is the result of an intentional ignition by a person or a naturally caused fire that is allowed to continue to burn according to approved plans to achieve resource management objectives

Vocabulary
Define the following terms using the space provided.

1. Aerial fuel:

2. Aspect:

3. Crown:

4. Cupping:

5. Duff:

6. Fire head:

7. Fire heel:

8. Fire storm:

9. Firing out:

10. Fire flanks:
___
___
___

11. Fulgurites:
___
___
___

Short Answer
Complete this section with short written answers, using the space provided.

1. Identify the four major wildfire fuel groups.
___
___
___
___
___
___

2. List the six primary factors that affect wildfire spread.
___
___
___
___
___
___

3. Describe the effect of slope aspect on wildfire spread.
___
___
___
___
___
___

4. List six visual indicators that may aid an investigator in determining the path of a wildfire spread.

5. List 10 standard firefighting safety orders that apply for wildfire incidents.

Fire Alarms

The following case scenario will give you an opportunity to explore the concerns associated with fire investigation. Read the scenario and then answer each question in detail.

You are investigating the origin and cause of a wildfire that destroyed several hundred acres of property, including many residential structures. You are getting conflicting information from witnesses and area residents in that there are several views and opinions on the direction of the fire travel and the location of origin.

1. What indicators would you be searching for to aid in determining fire spread direction?

___

2. How would you conduct a search for the origin of the fire?

___

Management of Complex Investigations

Workbook Activities

The following activities have been designed to help you. Your instructor may require you to complete some or all activities as a regular part of your investigator training program. You are encouraged to complete any activity your instructor does not assign as a way to enhance your learning in the classroom.

Chapter Review

The following exercises provide an opportunity to refresh your knowledge of this chapter.

Matching

Match each of the terms in the left column to the appropriate definition in the right column.

_____ 1. Interested party   **A.** An outline of the tasks to be completed and includes the order and timeline for completion

_____ 2. Complex investigation   **B.** A description of the specific procedures and methods by which a task or tasks are to be accomplished

_____ 3. Evidence custodian   **C.** Person who is responsible for managing all aspects of evidence control

_____ 4. Protocol   **D.** Generally includes multiple simultaneous investigations and involves a number of interested parties

_____ 5. Work plan   **E.** Should be held before starting the on-scene investigation to discuss safety considerations, address concerns, and set forth the ground rules for conducting activities

_____ 6. Preliminary meeting   **F.** Any person, entity, or organization, including their representatives, with statutory obligations or whose legal rights or interests may be affected by the investigation of a specific incident

_____ 7. Flow chart   **G.** Can provide a general outline for a more detailed work plan of each of the work plan activities

_____ 8. Site safety person   **H.** Has the authority to stop activities if safety becomes an issue and until the issue can be addressed

CHAPTER 25

Multiple Choice

Read each item carefully and then select the best response.

_____ 1. Who are companies, persons, or other entities that may have an interest in the outcome in the investigation?
   A. Interested parties
   B. Evidence custodians
   C. Site safety persons
   D. Contractors

_____ 2. Accusations of evidence altering or other indiscretions can be avoided by
   A. Pictures
   B. Preliminary meetings
   C. Joint investigations
   D. Work plans

_____ 3. When should safety considerations, concerns, and the ground rules for conducting activities be discussed?
   A. Work plans
   B. Joint investigations
   C. Media briefings
   D. Preliminary meetings

_____ 4. Who needs to ensure an investigation site or scene evaluation is conducted to identify potential safety issues and then to address those issues?
   A. Incident commander
   B. Manager of the complex investigation
   C. Law agency
   D. Property owner

_____ 5. Regular communications among interested parties may be conducted in various forms such as
   A. Regular meetings
   B. Web sites
   C. Bulletin boards
   D. All of the above

_____ 6. These allow interested parties to be informed during times when they are not required to be at the investigation site:
   A. Web sites
   B. Regular meetings
   C. Bulletin boards
   D. Sticky notes

_____ 7. The handling of objections to the investigation protocol should be addressed in
   A. The safety plan
   B. The preliminary meeting
   C. The protocol
   D. The contract

____ 8. Who should be allowed to document and see the evidence before it is changed or removed from the investigation site?
   A. The owner
   B. The suspect(s)
   C. All interested parties
   D. The IC

____ 9. How should all evidence collected at the investigation site be stored?
   A. Placed in evidence bags and left on scene
   B. Taken back to the firehouse
   C. Secured in the trunk of the sheriff's car
   D. Secured in a locked facility with fire protection features

____ 10. These types of issues generally dictate that all interested parties participate in any destructive examination of the evidence:
   A. Aeration
   B. Spoliation
   C. Logistical
   D. Documentation

Fill-in
Read each item carefully and then complete the statement by filling in the missing word(s).

1. Site and scene safety is in the interest of all _____ _____ and needs to be addressed early in the investigation and planning process.

2. A _____ _____ should be developed with the interested parties.

3. The _____ and work plan may be separate documents or combined in one document.

4. Joint investigations help to avoid accusations of _____ _____ or other indiscretions.

5. The _____ or _____ need to be flexible because it may be necessary to amend an agreement as the scene or activity progresses or changes.

6. Disagreements should be noted in writing with _____ _____ to handle the disagreement and a resolution provided.

7. The communication should be maintained to keep the parties informed of the _____ _____ _____ of the scene work and scene examination.

8. Good communication on _____ is important because if parties have good notice of the activities and the schedule they can arrange their own calendars to accommodate the investigation schedule.

Vocabulary
Define the following terms using the space provided.

1. Evidence custodian:
   _____
   _____
   _____

2. Interested party:

3. Protocol:

4. Work plan:

Short Answer
Complete this section with short written answers, using the space provided.

1. What are some of the concerns that should be addressed during the preliminary meetings?

2. Identify six types of costs that may be incurred during an investigation that could involve cost sharing between interested parties.

3. Identify and generate a generic list of potential interested parties.

___

4. Differentiate between an investigation protocol and an investigation work plan.

___

5. Identify four ways regular communications may be conducted between interested parties.

___

Fire Alarms

The following case scenario will give you an opportunity to explore the concerns associated with fire investigation. Read the scenario and then answer each question in detail.

You are the lead fire investigator of a mercantile structure fire. The 95,000-square-foot building was fully involved, and the entire contents—3.5 million dollars worth of product—was damaged beyond use.

1. Generate a list of interested parties in the investigation.

   _____
   _____
   _____
   _____
   _____

2. Develop a work plan for the investigation. Include the tasks to be completed and a timeline for completion.

   _____
   _____
   _____
   _____
   _____

Marine Fire Investigations

Workbook Activities

The following activities have been designed to help you. Your instructor may require you to complete some or all activities as a regular part of your investigator training program. You are encouraged to complete any activity your instructor does not assign as a way to enhance your learning in the classroom.

Chapter Review

The following exercises provide an opportunity to refresh your knowledge of this chapter.

Matching
Match each of the terms in the left column to the appropriate definition in the right column.

_____ 1. Vessel **A.** The outer skin of a vessel, including the bottom, sides, and main deck but not including the superstructure, masts, rigging, and other fittings.

_____ 2. Hatch **B.** Usually a continuous, horizontal division running the length of a vessel and extending athwart ships. This corresponds to floors in a building on land.

_____ 3. Bulkhead **C.** Electrical power supplied from shore via a cord set.

_____ 4. Bilge **D.** A broad grouping of every description of watercraft, other than a seaplane on the water, used or capable of being used as a means of transportation on the water.

_____ 5. Lazarette **E.** A component of a rolling element bearing that contains the elements and transfers the load to the bearing; generally used in pairs (inner and outer races).

_____ 6. Deck **F.** The stern cross-section of a square-sterned vessel

_____ 7. Races **G.** The vertical separations in a vessel that form compartments and that correspond to walls in a building. The term does not refer to the vertical sections of a vessel's hull.

_____ 8. Hull **H.** The lowest, interior part of a vessel's hull; the area where spilled water and oil can collect.

_____ 9. Transom **I.** Stowage compartment, often in the aft end of a vessel, sometimes used as a workshop.

_____ 10. Shore power **J.** An opening in a vessel's deck that is fitted with a watertight cover.

CHAPTER 26

Multiple Choice
Read each item carefully and then select the best response.

_____ 1. What is a protection feature for the safe operation of electronic equipment in explosive atmospheres and/or under irregular operating conditions?
   A. Shore power
   B. Intrinsically safe
   C. Superstructure
   D. Venturi

_____ 2. What is the space on a vessel designed for people to reside?
   A. Superstructure
   B. Cockpit
   C. Cabin
   D. Hold

_____ 3. A narrowed area within a carburetor that causes air to accelerate and creates a low-pressure area that draws fuel into the intake manifold of the engine:
   A. Shore power
   B. Intrinsically safe
   C. Superstructure
   D. Venturi

_____ 4. Most vapors from paints, solvents, and lubricants are
   A. Heavier than air
   B. Lighter than air
   C. Same vapor density as air
   D. Not a concern on vessels

_____ 5. The most common types of propulsion systems that the marine fire investigator will encounter are what type of power plants?
   A. Nuclear
   B. Gasoline or diesel engine
   C. Battery powered
   D. Solar

_____ 6. What types of electrical systems are normally found on marine vessels?
   A. AC
   B. DC
   C. Both AC and DC
   D. Neither AC nor DC

_____ 7. What are the vessel structures above the main deck or weather deck called?
   A. Bilge
   B. Transom
   C. Superstructure
   D. Venturi

_____ 8. A review of available statistical information for both large ship and recreational vessel fires suggests that between one-half and three-quarters of marine fires occur where?
   A. Engine spaces
   B. Galley, kitchen area
   C. In port
   D. The bilge

_____ 9. What is the most common electric propulsion unit the marine fire investigator will encounter?
   A. Lift truck
   B. Crane
   C. Bilge pump
   D. Trolling motor

_____ 10. If a pulley, alternator, or water pump seizes, what object can heat as it passes over these components, generating sufficient heat to ignite?
   A. Bilge
   B. Drive belt
   C. Races
   D. Vapors

Fill-in

Read each item carefully and then complete the statement by filling in the missing word(s).

1. Hydrogen gas may be located in battery compartments, and care should be taken to reduce the potential of _____ _____ or electrical arcs when removing the battery cables.

2. Marine vessels are often _____ to allow for containment of water should flooding occur.

3. On marine vessels that contain sewage holding tanks, _____ _____ _____ may be present.

4. The investigator must ensure that the craft is _____ before conducting the investigation because if the vessel lists or capsizes, the investigator may become trapped within it.

5. The fuels used in propulsion systems for marine vessels vary depending on the _____ _____ _____.

6. It is not uncommon to find both _____ _____ _____ systems on a vessel, especially those designed for continuous habitation.

7. In saltwater marine environments, a common difficulty facing vessel owners and operators is protecting electrical and mechanical components from _____.

8. Although _____ electrical items are usually protected from sea spray contact and saltwater intrusion, at times it still occurs.

9. The _____ and primary construction of smaller commercial vessels or recreational boats may be steel but may also consist of wood, fiberglass-reinforced plastic, aluminum, or ferro-cement.

10. Once fires involve spaces containing _____, they can easily grow out of control with little hope of firefighting.

Vocabulary

Define the following terms using the space provided.

1. Bilge:

2. Bulkhead:

3. Deck:

4. Hatch:

5. Hull:

6. Lazarette:

7. Races:

8. Venturi:

Short Answer

Complete this section with short written answers, using the space provided.

1. List 11 factors that can affect the interpretation of fire ignition, growth, and pattern development on a marine vessel.

2. Identify four potential ignition sources for vessel fires.

3. Identify the four areas that may be the area of origin in a vessel fire.

4. Detail the procedure for inspecting a marine fuel tank and hose fill system to determine if it had a role in supplying fuel to the fire.

5. List three types of marine vessels to which an investigator may have to respond.

Chapter 26 : Marine Fire Investigations

Labeling
Label the following exterior and interior marine vessel diagrams.

1. Basic marine vessel diagram.

   A. Exterior.

   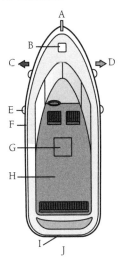
   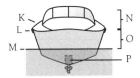

   B. Interior.

   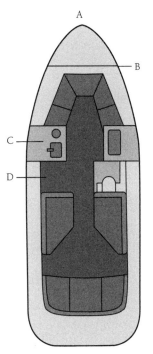

A. _____
B. _____
C. _____
D. _____
E. _____
F. _____
G. _____
H. _____
I. _____
J. _____
K. _____
L. _____
M. _____
N. _____
O. _____
P. _____

A. _____
B. _____
C. _____
D. _____

Fire Alarms

The following case scenario will give you an opportunity to explore the concerns associated with fire investigation. Read the scenario and then answer each question in detail.

You have been assigned to investigate a boat fire in the local marina. On arrival, you are led to a 32-foot recreational vessel that has sustained considerable fire damage but is still afloat and moored to the dock. The deck has burnt away in places and there appears to be several feet of water in the hull.

1. What is the greatest hazard this investigation poses to you, the investigator?

   _____
   _____
   _____
   _____
   _____

2. How would you conduct your witness interviews? What type of questions would you ask them?

   _____
   _____
   _____
   _____
   _____
   _____

Answer Key

Chapter 1: Administration

Matching

1. F (page 5)
2. H (page 5)
3. A (page 7)
4. I (page 4)
5. G (page 4)
6. B (page 4)
7. J (page 7)
8. C (page 7)
9. D (page 8)
10. E (page 8)

Multiple Choice

1. C (page 5)
2. D (page 5)
3. B (page 11)
4. A (page 11)
5. C (page 7)
6. D (page 7)
7. B (page 7)
8. B (page 7)
9. A (page 8)
10. D (page 8)

Fill-in

1. investigating; analyzing (page 4)
2. establish guidelines; recommendations (page 4)
3. NFPA 1033 (page 5)
4. codes; standards; recommended practices; guides (page 7)
5. technical (page 7)
6. 3; 5 (page 8)
7. call for proposals (page 8)

Vocabulary

1. **Guide:** A document that is advisory or informative in nature and that contains only nonmandatory provisions. A guide may contain mandatory statements, such as when a guide can be used, but the document as a whole is not suitable for adoption into law. (page 5)
2. **Standard:** A document, the main text of which contains only mandatory provisions using the word "shall" to indicate requirements and that is in a form generally suitable for mandatory reference by another standard or code or for adoption into law. Nonmandatory provisions shall be located in an appendix or annex, footnote, or fine print note and are not to be considered a part of the requirements of a standard. (page 5)
3. **Fire Investigation:** The process of determining the origin, cause, and development of a fire or explosion. (page 5)
4. **NFPA 921 Guide for Fire and Explosion Investigations:** A guide that establishes guidelines and recommendations for the safe and systematic investigation or analysis of fire and explosion incidents. (page 5)
5. **NFPA 1033 Professional Qualifications for Fire Investigator:** A standard designed to establish the minimum job performance requirements (JPRs) for service as a fire investigator. (page 5)

Short Answer

1. NFPA 1033 requires a fire investigator to be at least 18 years of age and possess a high school diploma or equivalent level of education. As with any profession, fire investigators must meet a basic level of education and training. The fire investigator must meet the JPRs established in NFPA 1033 and their employer as well as any requirements established by law. (page 5)
2. An up-to-date knowledge of the following topics should be maintained as a minimum for the fire investigator:
   - Fire science
   - Fire chemistry
   - Thermodynamics
   - Thermometry
   - Fire dynamics

- Explosion dynamics
- Computer fire modeling
- Fire investigation
- Fire analysis
- Fire investigation methodology
- Fire investigation technology
- Hazardous materials
- Failure analysis and analytical tools (page 7)

3. All NFPA documents are voluntary documents, which means they do not have the power of law unless an AHJ adopts them. (page 7)

4. A standard is a document in which the main text contains only mandatory provisions and is in a form that is generally suitable for adoption into law, assuming the AHJ adopts them. An example of a standard is NFPA 1033.

   A guide is a document that is advisory or informatory and that contains nonmandatory provisions. The document is not suitable for adoption into law. An example of a guide is NFPA 921. (page 11)

5. NFPA uses the following membership categories to fill and balance a committee:
   - Manufacturer
   - User
   - Installer/maintainer
   - Labor representative
   - Enforcing authority
   - Insurance representative
   - Special expert
   - Consumer
   - Applied research/testing laboratory (page 7)

Fire Alarms

1. The first step for any NFPA document entering its revision cycle occurs when a call for proposals is issued. An indication that a document is entering its revision cycle, along with the dates for accepting proposed changes, appears in NFPA publications, on the NFPA Web site (www.nfpa.org), and in various professional and governmental publications that serve parties interested in the subject matter. Anyone, whether a member of NFPA or not, can submit a proposal for the technical committee to consider during a 20-week window. The form for submitting proposals for change includes a section in which the proposer must provide substantiation for the change. The proposal forms can be found at the back of NFPA documents, including NFPA 921, or on the NFPA Web site. (page 8)

Chapter 2: Basic Fire Methodology

Matching

1. G (page 16)
2. E (page 17)
3. B (page 19)
4. A (page 16)
5. D (page 19)
6. C (page 20)
7. F (page 22)
8. H (page 20)

Multiple Choice

1. B (page 16)
2. C (page 16)
3. A (page 17)
4. B (page 17)
5. D (page 17)
6. C (page 17)
7. A (page 17)
8. C (page 17)
9. B (page 19)
10. B (page 19)

Chapter 2: Basic Fire Methodology

Fill-in

1. testing the hypothesis (page 19)
2. expectation bias (page 20)
3. technical review (page 20)
4. administrative review (page 20)
5. suspected; possible (page 19)
6. investigator (page 19)
7. deductive reasoning (page 19)
8. technical review (page 20)

Vocabulary

1. **Deductive reasoning:** The process by which conclusions are drawn by logical inference from given premises. Use of knowledge, skills, and art to challenge or test the hypothesis analytically. (page 19)
2. **Empirical data:** Data that are collected based on observation or experience and are capable of being verified. (page 17)
3. **Hypothesis:** Theory supported by the empirical data that the investigator has collected through observation and then developed into explanations for the event, which are based on the investigator's knowledge, training, experience, and expertise. (page 16)
4. **Scientific method:** The systematic pursuit of knowledge involving the recognition and formulation of a problem, the collection of data through observation and experiment, and the formulation and testing of a hypothesis. (page 16)

Short Answer

1. Probable: More likely true than not (more than 50% likely of being true).
   Possible: Feasible. Often used when two hypotheses have the same level of certainty.
   Suspected: Not enough certainty to be considered an expert opinion. (page 19)
2. The eight types of questions the investigator should use in testing the hypothesis are as follows:
   - Is there another way to interpret the facts (data)?
   - If yes, why is your interpretation more likely true?
   - What are the weaknesses in the hypothesis (analysis of the data)?
   - What arguments will someone else (an opposing expert) use to refute the hypothesis?
   - Are there facts that contradict the hypothesis?
   - Is there research that supports the hypothesis?
   - Can the hypothesis be proven to someone else?
   - Does the hypothesis make sense? (page 19)
3. The seven categories of data are as follows:
   - Recognition of physical evidence, such as fire patterns, fuel loads, ventilation, building construction, and other facts
   - Collection of materials, such as debris samples, for laboratory evaluation
   - Results of laboratory examinations
   - Documentation of personal observations, such as witness statements
   - Documentation of the fire scene through photographs, sketches, and notes
   - Official reports, such as those of fire and police departments
   - The documentation or results of prior scene investigations (page 17)
4. The steps that should be followed in using the scientific method for fire investigations are the following:
   - Receive the assignment: The investigator is notified of the fire loss and goal of the requested investigation.
   - Prepare for the investigation: For the given assignment and the nature of the fire scene, the investigator should determine the tools required, safety concerns to be addressed, and personnel needs to complete the task.
   - Conduct the investigation: Begin a systematic process of information gathering to include examination of the scene, interviews, and information research.
   - Collect and preserve evidence: Identify possible evidence and document and preserve it for future testing or legal presentation.
   - Analyze the incident: Using the scientific method, the collected and available data are analyzed. A hypothesis is developed and tested.

- Draw conclusions: A final tested hypothesis is determined.
- Reporting procedures: As determined by the responsibility of the investigator, the process and conclusions are reported in written or oral forms. (page 20)

5. These reviews are generally categorized into different types: administrative, technical, and peer reviews. (page 20)

Fire Alarms

1. The following steps should be followed in using the scientific method for fire investigations:
   1. Receive the assignment: The investigator is notified of the fire loss and goal of the requested investigation.
   2. Prepare for the investigation: For the given assignment and the nature of the fire scene, the investigator should determine the tools required, safety concerns to be addressed, and personnel needs to complete the task.
   3. Conduct the investigation: Begin a systematic process of information gathering to include examination and documentation of the scene, interviews, and information research.
   4. Collect and preserve evidence: Identify possible evidence and document and preserve it for future testing or legal presentation.
   5. Analyze the incident: Using the scientific method, the collected and available data are analyzed. A hypothesis is developed and tested.
   6. Draw conclusions: A final tested hypothesis is determined.
   7. Reporting procedures: As determined by the responsibility of the investigator, the process and conclusions are reported in written or oral forms. (page 20)

Chapter 3: Basic Fire Science

Matching

1. H (page 28)
2. J (page 29)
3. A (page 40)
4. E (page 29)
5. B (page 26)
6. G (page 31)
7. D (page 29)
8. C (page 29)
9. F (page 30)
10. I (page 29)

Multiple Choice

1. B (page 28)
2. C (page 26)
3. A (page 28)
4. D (page 26)
5. B (page 30)
6. D (page 26)
7. C (page 30)
8. D (page 29)
9. A (page 29)
10. D (page 29)
11. B (page 29)
12. D (page 29)
13. B (page 31)
14. C (page 33)
15. A (page 33)

Fill-in

1. thermal runaway (page 34)
2. oxidizing agent (page 35)
3. position; orientation (page 35)
4. melting; dripping (page 35)
5. fuel-controlled burning (page 36)
6. ventilation controlled (page 37)
7. fuel packages (page 37)
8. irregular patterns (page 38)
9. neutral plane (page 38)
10. increase (page 39)

Vocabulary

1. **Combustion:** A chemical process of oxidation that occurs at a rate fast enough to produce heat and usually light in the form of either a glow or flames. (page 26)
2. **Conduction:** Heat transfer to another body or within a body by direct contact. (page 29)
3. **Convection:** Heat transfer by circulation within a medium such as a gas or a liquid. (page 30)
4. **Flash point:** The temperature to which a liquid must be heated to sustain burning after the removal of an ignition source. Typically, only a few degrees higher than the flash point and in some instances the same. (page 33)

5. **Flashover:** Transition stage of a fire at which convected and radiated heat energy impinge on the other combustible items within the room, producing fire gases. These items then ignite at nearly the same time, causing full-room involvement. (page 37)
6. **Oxidizing agent:** A substance that promotes oxidation during the combustion process. (page 27)
7. **Pyrolysis:** Process in which something is heated, causing the material to decay and produce fire gases. (page 26)
8. **Radiation:** The combined process of emission, transmission, and absorption of energy traveling by electromagnetic wave propagation between a region of higher temperature and a region of lower temperature. (page 30)
9. **Thermal inertia:** Those properties of the material that characterize its rate of surface temperature rise when exposed to heat. It is the product of the thermal conductivity (κ), the density (p), and heat capacity (c). (page 29)
10. **Vaporization:** The process of producing ignitable vapors from a liquid. (page 26)

Short Answer

1. The most common fuels an investigator will encounter are organic fuels, which contain carbon. These fuels include wood, plastics, and petroleum products. (page 26)
2. The fourth component, an uninhibited chemical chain reaction, provides a self-sustaining event that continues to develop fuel vapors and sustain flames even after the removal of the ignition source. As a fire continues to burn, this exothermic reaction radiates heat back to the surface of the fuel, producing more vapors and continuing the combustion process. (page 28)
3. The "perfect" combustion condition is referred to as the stoichiometric ratio. The stoichiometric ratio is a concentration that exists above the lower explosive limit (LEL) and below the upper explosive limit (UEL). (page 28)
4. The color of smoke should not be relied on as an indicator of the material burning. Although certain fuels may produce a particular color and density of smoke, other factors (including decreasing oxygen levels and ventilation-controlled fires) can produce heavy, dark smoke commonly observed in oil and other hydrocarbons burning. As a fire goes through its various phases, smoke production generally increases. Firefighting operations can also change the color of smoke by mixing condensing vapors with black smoke, producing a white to gray color. (page 28)
5. The primary engine that creates flows in fires is the fire itself. The hot gases created by the fire rise above the fire in a plume. As these gases rise they entrain or draw in cool air, which causes the velocity of the gas flow to increase. At the same time the temperature of the plume decreases. The entrainment of air also causes the diameter of the plume to increase, resulting in a cone shape. When the plume reaches the ceiling of a room, the time it takes the fire gases to move parallel to the ceiling is known as a ceiling jet. The ceiling jet flows along the ceiling until it encounters a vertical obstruction, such as a wall, which impedes its flow. (page 29)

Fire Alarms

1. Questions you should ask the attack team and tip man include the following:
   - What ventilation tactics were in use before the initial fire attack, and how effective were they?
   - Were the door and room windows open, or did the attack team have to open them?
   - What was the estimated temperature of the room at ceiling level?
   - How much flame was visible at the chair and curtains? (pages 36–38)
2. The fire was extinguished in the pre-flashover stage. Answers to the questions from the tip man (#1 above) would help verify this as well as explain the lack of fire damage to the room and apartment. (page 37)

Additional Activity

1. Some common household oxidizing agents are
   - Pool chlorine
   - Hydrogen peroxide
   - Iodine
   - Nitric acid
   - Oxygen
   - Permanganate salts

2. Some common industrial oxidizing agents are
     - Organic peroxides
     - Inorganic peroxides
     - Fluorine
     - Chlorine
     - Bromine
     - Ammonium nitrate
     - Oxygen
     - Nitric acid

Chapter 4: Fire Patterns

Matching

  1. H (page 45)
  2. D (page 44)
  3. G (page 46)
  4. I (page 48)
  5. B (page 44)
  6. J (page 54)
  7. A (page 46)
  8. E (page 56)
  9. C (page 48)
  10. F (page 46)

Multiple Choice

  1. C (page 44)
  2. A (page 45)
  3. D (page 45)
  4. B (page 46)
  5. A (page 48)
  6. D (page 48)
  7. C (page 48)
  8. D (page 50)
  9. B (page 54)
  10. A (page 54)

Fill-in

  1. fire patterns (page 44)
  2. 1900 (page 45)
  3. Demarcation lines (page 46)
  4. expand (page 47)
  5. steel structures (page 47)
  6. insufficient (page 48)
  7. furniture springs (page 49)
  8. Heat shadowing (page 49)

Vocabulary

  1. **Beveling:** A fire pattern that indicates fire direction on wood wall studs. The bevel leans toward the direction of travel. (page 54)
  2. **Calcination:** Process in which chemically bound water is driven out of gypsum by the heat of the fire. (page 48)
  3. **Crazing:** Cracks that can be either straight or crescent shaped and can extend through the entire thickness of the glass. (pages 48–49)
  4. **Fire effects:** The observable or measurable changes in or on a material as a result of exposure to the fire. (page 44)
  5. **Heat shadowing:** A discontinuous pattern on a surface that indicates an interruption of heat transfer. (page 49)
  6. **Rainbow effect:** A diffraction pattern formed when hydrocarbons float on a surface. (page 50)
  7. **Spalling:** The chipping or pitting of concrete or masonry surfaces. (page 46)

Short Answer

  1. Many factors may affect the rate at which wood may char. Such conditions include
     - Rate and duration of heating
     - Ventilation effects
     - Surface-to-mass ratio
     - Direction, orientation, and size of wood grain
     - Species of wood (e.g., maple vs. yellow pine)
     - Moisture content of the wood product
     - Any surface coating on the wood (page 45)

2. Color changes are not without their limits. Not all people see color with the same perspective. Factors such as lighting conditions, angle of view, and nature of color are all factors to be considered. In addition, coloration changes may occur from non-fire factors such as exposure to sunlight. Fabric dyes may change in appearance related to heat exposure. How a particular dye may react to heat exposure may vary with the product. (page 46)

3. When exposed to temperatures in excess of approximately 538°C (1000°F), steel structures such as columns and beams begin to buckle and bend, ultimately failing. As steel is exposed to higher temperatures, the strength and load-carrying capability decreases until deformation of the material occurs. This deformation is the result of the object's inability to support the load placed on it. The load may be the object itself. This should not be confused with melting because the object has not liquefied, merely distorted. The distortion or deformity of an object indicates that the melting temperature was never reached. (page 47)

4. Hydrocarbons do not mix with water and are found floating on the surface, creating an interference pattern that produces a rainbow effect. This effect is present at fire scenes and should not be relied on as an indicator of the presence of an ignitable liquid. Many materials such as asphalt, plastic, and wood products can produce rainbow effects as a result of pyrolysis. (page 50)

5. Fire-suppression actions may create or change fire patterns present. Water streams may change the direction of fire spread, and ventilation actions during the fire growth affect the fire patterns present. The actions of individuals at the scene of the fire should be learned to understand whether the actions taken may have caused changes to patterns noted. (page 52)

Fire Alarms

1. The direction of fire travel through a horizontal surface is determined by examining the sides of the hole and the slope created by the fire. When the hole is wide and slopes downward from the top surface, the direction of fire travel is from above. In the case of a fire advancing from below a surface, the sides are wider on the bottom side and slope upward. Investigators should be cautious when determining the direction of fire travel through these penetrations because fire movement may have occurred in both directions, leaving only the last direction of travel present through the hole. (page 54)

Chapter 5: Building Systems

Matching

1. C (page 67)
2. F (page 64)
3. E (page 62)
4. A (page 63)
5. G (page 66)
6. I (page 65)
7. J (page 66)
8. B (page 65)
9. D (page 68)
10. H (page 67)

Multiple Choice

1. A (page 66)
2. A (page 68)
3. C (page 68)
4. C (page 70)
5. B (page 70)
6. A (page 70)
7. D (page 71)
8. D (page 71)
9. B (page 73)
10. C (page 75)

Fill-in

1. building construction (page 62)
2. interstitial spaces (page 62)
3. HVAC system (page 63)
4. Dead loads (page 64)
5. load-bearing (page 65)
6. occupancy classification (page 66)
7. Manufactured housing (page 69)
8. gypsum board (page 70)

Vocabulary

1. **Compartmentation:** A concept in which fire is confined to the room of origin, minimizing smoke movement to other areas of a building. (page 62)
2. **Dead load:** The weight of materials that are part of a building, such as the structural components, roof coverings, and mechanical equipment. (page 64)

3. **Interstitial spaces:** The space between the building frame and interior walls and the exterior facade and with spaces between ceilings and the bottom face to the floor or deck above. (page 62)
4. **Live load:** The weight of temporary loads that need to be designed into the weight-carrying capacity of the structure, such as furniture, furnishings, equipment, machinery, snow, and rain water. (page 63)
5. **Mill construction:** An early form of heavy timber construction influenced and developed largely through insurance companies that recognized a need to reduce large fire losses in factories. (page 70)
6. **Smoke barrier:** Continuous membrane, either vertical or horizontal, designed and constructed to restrict movement of smoke. (page 73)
7. **Soffits:** The horizontal undersides of the eaves or cornice. (page 68)

Short Answer

1. Additional loads applied above and beyond the structure's design parameters may create instability. Examples of such conditions include the following:
   - Extreme snow loads
   - Extreme wind loads
   - Additional contents, such as stock
   - Additional mechanical components, such as HVAC and elevator equipment
   - Large congregation of people in a limited area
   - Water from firefighting operations (page 64)

2. In compartment fires the following factors significantly affect fire spread:
   - Room size
   - Interior finish
   - Shape
   - Ceiling height
   - Placement and area of doors and windows
   - HVAC systems (page 63)

3. Laminated beams are structural elements that have the same characteristics as solid wood beams. They are composed of many wood planks that are glued or laminated together to form one solid beam and are generally for interior use only. They are commonly referred to as glulam beams. They behave like heavy timber until failure. Effects of weathering decrease their load-bearing ability. The investigator should document the size of individual members as well as the overall size of the beam.

   Wood I-beams have a smaller dimension than floor joists and therefore can burn through and fail sooner than dimensional lumber. Openings in the web for utilities may reduce the structural integrity of the web. I-beams must be protected by gypsum board to help delay collapse under fire conditions.

   Wood trusses are similar to other trusses in design. Individual members are fastened using nails, staples, or metal gusset plates (gang nail plates) or wooden gusset plates. Truss failure can occur from gusset plates failing even before wood members are burned through. When one member of the truss fails, the other members take on additional loads and may become stressed and cause the entire truss to fail. (page 70)

4. Steel: Can conduct electricity. Good conductor of heat. Loses its ability to carry a load well below temperatures encountered in a fire. Can distort, buckle, or collapse as a result of fire exposure. Amount of distortion depends on factors such as heat of the fire, duration of exposure, physical configuration, and composition of the steel.

   Masonry and concrete: Due to mass will generally absorb more heat than steel. Good thermal insulators. Do not heat up quickly and do not transfer through them as easily as steel. Masonry and concrete will carry a load much longer at equal temperatures when compared with steel. (page 71)

5. Factors that can affect the failure of floor, ceiling, and roof assemblies include type of structural element, protection from the elements, span, load, and beam spacing. Added live loads, such as water during firefighting operations, can contribute to failure. (page 71)

Labeling

1. Typical single-family dwelling (page 64).

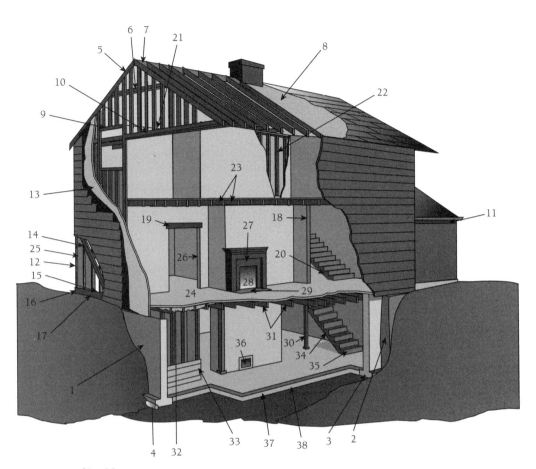

1. Foundation wall
2. Frost wall
3. Wall footing
4. Perimeter drain
5. Rafter
6. Collar beam
7. Ridge board
8. Roof sheathing
9. Window header
10. Attic joist
11. Box beam
12. Exterior wall stud
13. Wall sheathing
14. Corner bracing
15. Exterior wall plate
16. Box sill
17. Sill
18. Wall stinger
19. Header
20. Stair partition casing
21. Attic insulation
22. Partition studs
23. Second floor joists
24. Finish flooring
25. Wall insulation
26. Cripple stud
27. Damper control
28. Ash door
29. Hearth
30. Post or column
31. First floor joist
32. Subfloor
33. Basement partition
34. Stair stringer
35. Tread and riser
36. Cleanout door
37. Concrete floor slab
38. Granual fill

2. Types of building construction.

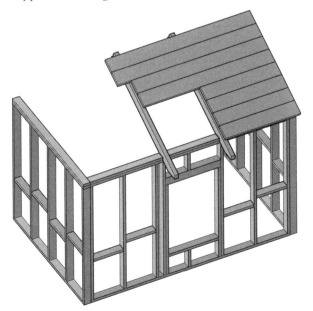

A. Wood frame construction (page 67)

180 ANSWER KEY

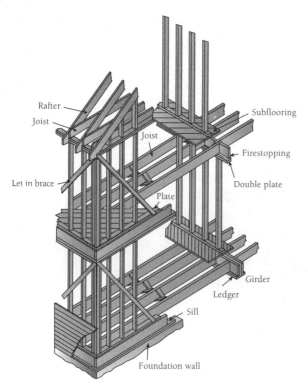

B. Platform frame construction (page 67)

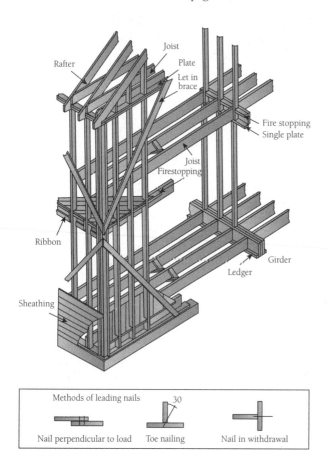

C. Balloon frame construction (page 68)

Chapter 5: Building Systems 181

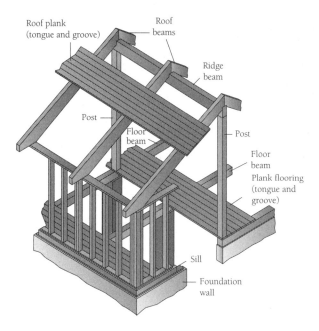

D. Plank-and-beam construction (page 68)

E. Post-and-frame construction (page 69)

F. Heavy timber construction (page 69)

Fire Alarms

1. Laminated beams are structural elements that have the same characteristics as solid wood beams. They are composed of many wood planks that are glued or laminated together to form one solid beam and are generally for interior use only.

   They are commonly referred to as glulam beams. They behave like heavy timber until failure. Effects of weathering decrease their load-bearing ability. The investigator should document the size of individual members as well as the overall size of the beam.

   Wood I-beams have a smaller dimension than floor joists and therefore can burn through and fail sooner than dimensional lumber. Openings in the web for utilities may reduce the structural integrity of the web. I-beams must be protected by gypsum board to help delay collapse under fire conditions.

   Lightweight wood trusses are similar to other trusses in design. Individual members are fastened using nails, staples, glue, or metal gusset plates (gang nail plates) or wooden gusset plates. Truss failure can occur from gusset plates failing even before wood members are burned through. When one member of the truss fails, the other members take on additional loads and may become stressed and cause the entire truss to fail. (page 70)

2. Safety procedures will have to be modified and strictly enforced because of the lack of structural integrity after fire damage.

   Damage to the structure occurs quicker relative to fire damage in older construction. This may cause an inexperienced investigator to believe a fire was burning longer than it actually had been, giving a false ignition time. (page 70)

Additional Activity

1. Fire fighters are at risk of falling through fire-damaged floors. Fire burning underneath floors can significantly degrade the floor system with little indication to fire fighters working above. Floors can fail within minutes of fire exposure, and new construction technology such as engineered wood floor joists may fail sooner than traditional construction methods. NIOSH recommends that fire fighters use extreme caution when entering any structure that may have fire burning beneath the floor.

   NIOSH lists several controls and recommendations in Publication No. 2009-114, *Preventing Deaths and Injuries of Fire Fighters Working Above Fire-Damaged Floors*, as follows:

   - Conduct a thorough fire size-up and communicate the findings to all personnel on-scene before entering the building. Incident commanders and company officers should be trained and experienced in structure fire size-up to avoid putting fire fighters at unneeded risk of working above fire-damaged floors.
   - Do not enter a structure, room, or area when fire is suspected to be directly beneath the floor or area where fire fighters would be operating, or if the location of the fire is unknown.
   - Never assume structural safety of any floor (regardless of the construction) having a significant fire under it.
   - Conduct preincident planning inspections during the construction phase to identify the type of floor construction. If preplanning is not conducted, assume residential construction and small commercial buildings built since the early 1990s may contain engineered wood I-joists.
   - Report construction deficiencies noted during preplanning to local building code officials. For example, engineered wood floor joists should only be modified per manufacturer specifications—usually limited to cutting to length and removing precut knockouts for utility access. Report damaged or cut chords or webs to building officials.

- Develop, enforce, and follow standard operating procedures on how to size up and combat fires safely in buildings of all construction types. Rapid intervention teams (RIT) should include a portable ladder with their RIT equipment when deployed at basement fires.
- Provide training on identifying signs of weakened floor systems (soft or spongy feel, heat transmitted through floor, downward bowing, etc.). Make fire fighters aware that all floor types can fail with little or no warning.
- Use a thermal imaging camera to help locate fires burning below or within floor systems, but recognize that the camera cannot be relied on to assess the strength or safety of the floor. Fire fighters should be trained on the use of thermal imaging cameras, including limitations and difficulties in detecting fire burning below floor systems.
- Immediately evacuate and, if possible, use alternate exit routes when floor systems directly beneath the floor where fire fighters would be operating are weakened by fire.
- Use defensive overhaul procedures after fire extinguishment in structures containing fire-damaged floor systems of all types.
- Consider becoming active in the building code process and influence requirements for fire resistance of floor and ceiling systems to further fire fighter safety and health.

In addition, NIOSH recommends the following:

- Trade associations and building contractors should consider providing education and training to fire service organizations on the hazards fire fighters face when fighting fires that have weakened all types of structural members. An example of such training is available at www.woodaware.info.
- Builders, contractors, and owners should consider protecting all floor systems, including engineered wood I-joists, by covering the underside with fire-resistant materials (Underwriters Laboratories, 2008).
- Builders, contractors, and owners should consider incorporating sprinkler systems into residential construction. Sprinkler use reduces the chances of both residential and fire fighter fatalities (USFA, 2008).

Chapter 6: Electricity and Fire

Matching

1. D (page 83)
2. E (page 85)
3. G (page 83)
4. J (page 84)
5. F (page 87)
6. I (page 87)
7. B (page 86)
8. C (page 86)
9. A (page 84)
10. H (page 85)

Multiple Choice

1. A (page 100)
2. D (page 89)
3. D (page 96)
4. B (page 90)
5. B (page 81)
6. A (page 96)
7. B (page 92)
8. B (page 89)
9. A (page 95)
10. D (page 85)
11. D (page 84)
12. C (page 92)
13. D (page 91)
14. A (page 101)
15. C (page 90)

Fill-in

1. resistance (page 81)
2. ampere rating (page 83)
3. service entrance (page 86)
4. ground connection (page 87)
5. Circuit breakers (page 89)
6. circuits (page 90)
7. ampacity (page 91)
8. eutectic melting (page 91)
9. AWG sizes (page 91)
10. GFCI (page 92)

Vocabulary

1. **Arc-fault circuit interrupter (AFCI):** Designed to protect against fires caused by arcing faults in home electrical wiring. The circuitry continuously monitors current flow. (page 90)
2. **Eutectic melting:** Any combination of metals with a melting point lower than that of any of the individual metals of which it consists. (page 91)

3. **Hot legs:** The two insulated conductors of a single-phase system (or three if three-phase power is used), sometimes referred to as L1, L2, or Line voltages. (page 85)
4. **Overfusing:** A dangerous condition that occurs when the circuit protection (fuse or circuit breaker) rating significantly exceeds the ampacity of the conductor, leading to a condition in which increased heat can occur in the conductors. (page 91)
5. **Resistance heating:** Occurs when current flows through a path that provides high resistance to current flow, such as a heating element (intentional) or a resistive connection (unintentional). (page 94)
6. **Resistive circuit:** A circuit that does not contain inductance and capacitance. (page 81)
7. **Sine wave:** The waveform that AC voltage follows. An example is 120 V AC, which has 170 V peak, crosses 0 to –170 V, and repeats this cycle 60 times per second. (page 84)
8. **Thermistors:** Sensors used to measure temperature by change of their resistance. (page 96)
9. **Voltage sniffer:** A noncontact voltage monitor that outputs a beep or turns on a light when voltage is present or nearby (within about an inch or less). (page 92)
10. **Weatherhead:** The point where service entrance cables connect to the structure, which is designed to keep water out of the conduit that carries the wires. (page 86)

Short Answer

1.

Hydraulics	Electricity
A **pump** creates the force that moves the water.	A **generator/battery** creates the electromotive force that moves the electrons.
Pressure is measured in pounds per square inch (psi) and is measured with a pressure gauge.	**Voltage (E)** is measured in volts (V) and is measured with a voltmeter.
Water moves through the pipes and does the work.	**Electrons** move through the conductors and do the work.
Flow is measured in gallons per minute (gpm) and is measured with a flow meter.	**Current (I)** is measured in amperes or amps (A) and is measured with an ammeter.
A **valve** controls the flow of water: • Open: water flowing • Closed: no water flowing	A **switch** controls the flow of electricity: • Off: open circuit; no electricity flowing • On: closed circuit; electricity flowing
Friction is the resistance of the pipe or hose to the water moving through it and is measured in psi.	**Resistance (R)** is the opposition of the conductors to the electrons moving through it and is measured in ohms (Ω).
Friction loss is the amount of pressure lost between two points in a pipe layout.	**Voltage drop** is the amount of voltage drop between two points in a circuit.
Pipe or hose size is measured in inches, inside diameter: • Larger pipe is greater flow • Smaller pipe is lower flow	**Conductor size** is given in AWG or wire gauge size: • Larger wire is greater current • Smaller wire is lower current

(page 81)

2. The three most commonly used conductors are copper, aluminum, and copper-clad aluminum conductors. (page 91)
3. Insulation on the conductors prevents faulting or leakage via unwanted paths. Insulation includes any material that can be applied readily to conductors, does not conduct electricity, and retains its properties for an extended time at elevated temperatures. Air can be an insulator, even for high voltages, when the conductors are kept separate. An arc can occur when dust, pollution, or products of combustion contaminate air or the insulation. The conductor's applications and insulation ratings are marked onto the insulation to identify its type and list its temperature rating.

Polyvinyl chloride, or PVC, a common insulator, can become brittle with age or heat and may give off hydrogen chloride in a fire. Rubber was a common insulating material until the 1950s. It can become brittle with age, and once it becomes brittle it can easily be broken off the conductor. Rubber chars and leaves an ash when exposed to fire. Other materials used as conductors include polyethylene and related polyolefins. These materials are commonly used on higher current circuits in residential applications. Nylon jackets may be used to increase thermal stability. (page 91)

4. The following conditions must exist for ignition from an electrical source:
   - Wiring must be charged with energy from some source (power must be on).
   - Sufficient heat and temperature must be produced (enough energy to ignite the first fuel; see Chapter 16, Fire Cause Determination).
   - Combustible material must be a material that can be ignited by the energy produced by the electrical failure (to raise temperature to ignite).
   - The heat source and the combustible fuel must be close enough for a sufficient period of time for the combustible material to generate combustible vapors; a simple arc does not produce enough energy to ignite most ordinary combustibles (page 92)

5. Some methods that may generate sufficient heat include the following:
   - Resistance heating that occurs when current flows through something that provides high resistance to current flow, such as a heating element (intentional) or a resistive connection (unintentional).
   - Ground fault or equipment return path short circuits in which excessive current flows in the return path to the source instead of the intended circuit path.
   - Parting arcs that occur when wires or switch contacts are separated and current flow is interrupted. This creates a momentary arc of very short duration, with accompanying sparks.
   - Excessive current, as when there is more current flowing than an appliance or conductor is intended to carry and may overheat (e.g., carrying 50 instead of 8 A).
   - Light bulb over lamping, close proximity of combustibles to heaters, or cooking equipment.
   - Heating in a confined space such as a heating element that is normally exposed to ambient or cooling air but that is somehow trapped under blankets or bunched up. Another example of confined heating is a heat-producing object such as a motor, which is covered in dust or is located in an insulated box (page 92)

Labeling

1. Triplex overhead service drop.

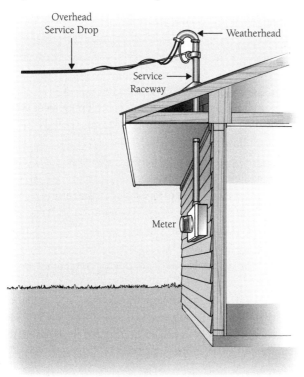

Fire Alarms

1. It is important for the investigator to understand the relationship among voltage, current, and resistance and how to calculate power from these. This knowledge will help the investigator understand the potential of electricity to be a fire cause. For example, knowing voltage and resistance, or voltage and power, or power and resistance will allow the investigator to determine how much current or energy was used by an appliance, whether a circuit was overloaded, or whether the overcurrent protection was properly matched to the power requirements for the circuit. The Ohm's law wheel is useful for determining one parameter in terms of any two others.

   The investigator should be able to calculate the total power requirements of a circuit with multiple loads on it by inspecting the equipment and determining the power requirements of each piece of equipment and then totaling all these values. The nameplate information on the appliances provides some information regarding important electrical parameters. Often, these values are given in amps, watts, or voltage. Keep in mind these are typical maximum values (the upper end of nominal state voltage) and assume active operation of the equipment during peak power utilization. Obviously, a washing machine is not going to draw full rated current unless it is in the agitation or spin cycle, where the power required to overcome inertia is highest. Because power use is dynamic and can ebb and flow, branch circuit overload situations can develop with multiple different loads operating intermittently, if they occur concurrently. Electrical system capacity is not designed for all loads running at maximum power use all the time, but during times of abnormally high circuit and appliance utilization, power use can exceed circuit capacity, and circuit breakers may open to protect the circuits.

   By using Ohm's law, we see that when a fault occurs in a circuit, the circuit resistance decreases, and the current increases to an abnormally high value; however, if sufficient resistance exists in the electrical current path, the short circuit current may continue. Thus, even though the resistance of a fault in a circuit might be low, it can still be high enough to dissipate significant energy while drawing lower amperage than is required for the circuit protection device to trip. For example, if a fault develops across a carbon path between two conductors and there is 100 ohms of resistance, the current draw for that particular fault would be 1.2 amps (amps is volts/ohms), and the circuit protection may not trip; however, if a fault occurs where there is very little resistance, such as 0.2 ohms, there will be up to 600 amps flowing through that fault, and the circuit protection should immediately open. In the former case, more than 100 watts is dissipated in the fault path, and this can generate sufficient heat to ignite nearby combustible material. (page 82)

2. The following conditions must exist for ignition from an electrical source:
   - Wiring must be charged with energy from some source (power must be on).
   - Sufficient heat and temperature must be produced (enough energy to ignite the first fuel; see Chapter 16, Fire Cause Determination).
   - Combustible material must be a material that can be ignited by the energy produced by the electrical failure (to raise temperature to ignite).
   - The heat source and the combustible fuel must be close enough for a sufficient period of time for the combustible material to generate combustible vapors; a simple arc does not produce enough energy to ignite most ordinary combustibles.

   Some methods that may generate sufficient heat include the following:
   - Resistance heating that occurs when current flows through something that provides high resistance to current flow, such as a heating element (intentional) or a resistive connection (unintentional).
   - Ground fault or equipment return path short circuits in which excessive current flows in the return path to the source instead of the intended circuit path.
   - Parting arcs that occur when wires or switch contacts are separated and current flow is interrupted. This creates a momentary arc of very short duration, with accompanying sparks.
   - Excessive current, as when there is more current flowing than an appliance or conductor is intended to carry and may overheat (e.g., carrying 50 instead of 8 A).
   - Light bulb over lamping, close proximity of combustibles to heaters, or cooking equipment.
   - Heating in a confined space such as a heating element that is normally exposed to ambient or cooling air but that is somehow trapped under blankets or bunched up. Another example of confined heating is a heat-producing object such as a motor, which is covered in dust or is located in an insulated box. (page 92)

Additional Activity

1. Ohm's law is voltage = current × resistance. Using Ohm's law, calculate the following:
   - **A.** Voltage = 120 volts (10 × 12 = 120)
   - **B.** Voltage = 250 volts (25 × 10 = 250)
   - **C.** Voltage = 440 volts (50 × 8.8 = 440)
   - **D.** Current = 11 amps (110 / 10 = 11)
   - **E.** Current = 11 amps (220 / 20 = 11)
   - **F.** Current = 88 amps (440 / 5 = 88)
   - **G.** Resistance = 11 ohms (110 / 10 = 11)
   - **H.** Resistance = 44 ohms (220 / 5 = 44)
   - **I.** Resistance = 11 ohms (440 / 40 = 11)

Chapter 7: Fuel Gas Systems

Matching

1. C (page 111)
2. I (page 124)
3. D (page 113)
4. H (page 111)
5. G (page 115)
6. J (page 124)
7. F (page 116)
8. A (page 124)
9. E (page 113)
10. B (page 113)

Multiple Choice

1. B (page 110)
2. A (page 111)
3. D (page 112)
4. B (page 112)
5. C (page 112)
6. D (page 113)
7. A (page 113)
8. B (page 113)
9. C (page 113)
10. D (page 115)
11. B (page 121)
12. A (page 121)

Fill-in

1. fugitive gases (page 111)
2. Fuel gases (page 111)
3. fusible plugs (page 113)
4. Vaporizers (page 115))
5. diaphragm type; lever type (page 115)
6. electronic arc (page 116)
7. bonded; grounded (page 118)
8. venting (page 118)
9. systematic manner (page 119)
10. pilot light (page 120)

Vocabulary

1. **Backflow valve:** Prevents gas from reentering a container or distribution system. (pages 113–114)
2. **Container appurtenances:** Devices connected to the openings in tanks and other containers—such as pressure relief devices, control valves, and gauges. (page 113)
3. **Fixed level gauge:** Primarily used to indicate when the filling of a tank or cylinder has reached its maximum allowable fill volume. They do not indicate liquid levels above or below their fixed lengths. (page 115)
4. **Lower explosive limit (LEL):** The lowest concentrations of fuel in a specified oxidant; also known as the lower flammable limits (LFLs). (page 112)
5. **Vaporizers:** Heaters used to heat and vaporize propane, where larger quantities of propane are required, such as for industrial application. (page 115)
6. **Variable gauge:** Gauges give readings of the liquid contents of containers, primarily tanks or large cylinders. They give readings at virtually any level of liquid volume. (page 115)

Short Answer

1.

	Natural Gas	Propane
Composition	Hydrocarbon gas Primary methane	95 percent propane and propylene 5 percent other gases
Density	Lighter than air Vapor density of 0.59–0.72	Heavier than air Vapor density of 1.5–2.0
LEL	3.9–4.5 percent	2.15 percent
UEL	14.5–15 percent	9.6 percent
Ignition temperature	900–1170°F (483–632°C)	920–1120°F (493–604°C)
BTUs (per cubic foot)	1030	2490
Delivery	Delivered via a distribution system directly to the customer	Stored in tanks on the customer's property and delivered via tank trucks

(page 111)

2. Natural gas is supplied through a transmission pipeline, then through main pipelines (mains), and then through service mains or service laterals before it is connected to the meter at the consumer's location.

    LPG distribution systems are similar in operation to natural gas systems except that the LPG storage supply is often located at the consumer's site. LPG storage containers can also be housed in bulk storage locations and piped underground, similar to natural gas systems. (page 112)

3. A tank is defined as a storage container with greater than 1000-lb (454 kg) water capacity or 800-lb (363 kg) LPG capacity. The design and construction of LP storage tanks are governed by regulations of the American Society of Mechanical Engineers (ASME). The typical working pressure in the storage tanks varies depending on the temperature of the LP but can range as high as 200 to 250 psi (1379–1724 kPa).

    Cargo tanks are those containers permanently mounted on a chassis and are used for transporting LPG. Portable tanks are used for transporting LPG, but they are not mounted on a chassis. Their quantities exceed a 454-kg (1000-lb) water capacity.

    Cylinders are considered upright containers. They have a water capacity of 1000 lb (454 kg) or less. The design and construction of cylinders are governed by regulations of the U.S. Department of Transportation. The pressure in the cylinders can be the same as that in tanks or containers, ranging up to 200 to 250 psi (1379–1724 kPa). Cylinders are most frequently used in rural homes and businesses, mobile homes, and recreational vehicles and for outdoor barbecue grills and motor fuel. (page 112)

4. Pressure relief valves are designed to open at a specific pressure, usually around 250 psi (1724 kPa). The pressure relief valve is generally placed in the container where it releases the vapor, although there are some exceptions. The pressure in a container is directly related to the temperature of the LPG. As the temperature rises, the pressure continues to rise. NFPA 58, *Liquefied Petroleum Gas Code*, provides methods to determine the pressure in a container if the temperature of the gas is known.

    Fusible plugs are thermally activated devices that open and vent the contents of a container. Once activated, they cannot be reused. Above-ground storage tanks with less than 1200 lbs (544 kg) of water capacity may have fusible plugs. The melting temperature of these plugs is between 208°F (98°C) and 220°F (104°C). (page 113)

5. The most common types of pressure regulators are the diaphragm type, which has a spring that is set to control the pressure, and the lever type. The amount of gas used may also be measured through a pressure regulator commonly known as a gas meter. The vents on the regulators must be clear for the device to operate properly. If the vent becomes plugged or obstructed, the pressure regulator device might not function properly. In cold or flood-prone environments, it is important not to place regulators where water or ice accumulations cannot obstruct the vent openings. (page 115)

Labeling

1. Water heater components.

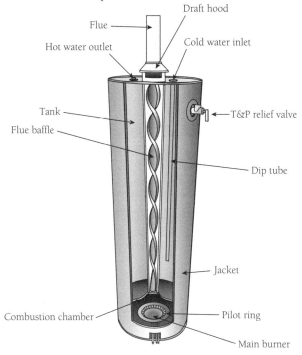

2. Furnace components.

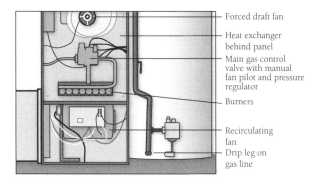

Fire Alarms

1. Analysis of the fuel gas system can provide information if it was involved in the origin or cause of the explosion or fire. As with all parts of a fire investigation, the analysis of the fuel gas system and each component of that system should be done in a systematic manner to ensure thoroughness. The goal is to determine whether and to what extent it operated or failed. NFPA 921 provides guidance in conducting the investigation. The investigator should interview all witnesses, property owners, fire fighters, and so forth to determine what condition and activities occurred before the fire or explosion that may have involved the fuel gas system of the building. Certain activities such as moving equipment, servicing appliances, and the installation of new fixtures may be factors that the investigator should consider.

The main causes of gas-fueled fires and explosions involve leakage in the fuel gas delivery system or in the appliance itself. The most common leakage occurs at pipe junctions, unlit pilot lights or burners, uncapped pipes, malfunctioning appliances and controls, areas of corrosion in pipes, and points of physical pipe damage.

It is possible for gases to migrate great distances underground before entering a structure or exiting the ground. The gas can enter underground gas lines, underground sewer lines, underground electrical and telephone conduits, or underground drain tiles. Gas can enter structures by migrating through cracks or holes in foundations. Weather changes (such as frozen ground) can cause the gas to be routed in a different direction or to migrate underground for greater distances.

When gas passes through a filter medium such as soil, the medium can "scrub" the odorant out of the gas. The gas then becomes odorless and might not be detected. NFPA 54 requires that gas meters be installed at least 3 feet (0.9 meters) from sources of ignition and protected from damage. Odorant can also be lost through adsorption into the piping or container system. This can occur with both new steel or plastic piping or containers or components that have been in use for an extended period of time. Loss of odorant can also occur as a result of reduced flow rates during seasonal changes as usage levels decrease. Odorant levels can also be decreased when mercaptan odorants oxidize as a result of contact with ferrous-metal containers. Odorant may also be trapped in water or other liquids that may collect within the container or piping system as a result of contamination or condensation caused by their ability to be soluble in water. (pages 119–122)

2. If the investigator suspects that fuel gas may be a factor in the fire investigation, he or she should test for leaks of the gas supply system before its disassembly. Expertise in conducting such evaluations is needed. (page 121)

Additional Activity

1. **A.** Table 7-1 lists the lower explosive limit for natural gas as 3.9 to 4.5 percent and the vapor density as .59 to .72. A CGI reading of 1.5 percent is below the lower explosive limit, indicating you do not have enough fuel in the atmosphere at the leak to support ignition. Natural gas has a vapor density less than air (1), indicating the gas will rise.

   B. Keep in mind natural gas may pocket in the ceiling area and accumulate to a concentration at or above its lower explosive limit, creating an explosive atmosphere away from the leak. Ceilings are notorious for having ignition sources. You may have a very dangerous atmosphere in a location away from the leak.

Chapter 8: Fire-Related Human Behavior

Matching

1. D (page 128)
2. G (page 129)
3. E (page 129)
4. A (page 128)
5. F (page 130)
6. J (page 133)
7. H (page 133)
8. B (page 129)
9. C (page 132)
10. I (page 132)

Multiple Choice

1. A (page 128)
2. B (page 128)
3. C (page 128)
4. B (page 129)
5. D (page 129)
6. C (page 129)
7. A (page 129)
8. C (page 129)
9. B (page 130)
10. D (page 133)

Fill-in

1. physiological factors (page 128)
2. familiarity (page 128)
3. size (page 129)
4. response times (page 129)
5. threats (page 129)
6. safety (page 130)
7. irregularities (page 132)
8. statement of danger (page 132)

Vocabulary

1. **Adolescent fire-setters:** Adolescents (ages 14–16) who are often responsible for fires that occur at places other than their homes. (page 133)
2. **Child fire-setters:** Children (ages 2–6) who are often responsible for fires in their homes or in the immediate area. (page 133)

3. **Juvenile fire-setters:** Juveniles (ages 7–13) who are often responsible for fires that start in their homes or in the immediate environment. (page 133)
4. **Statement of the danger:** A warning that identifies the nature and extent of the danger and the gravity of the risk of injury. (page 132)

Short Answer

1. An individual's familiarity with the setting can make escape more likely, although physical limitations and cognitive impairments can minimize the advantages of such familiarity, such as a person's becoming lost in his or her own home. In larger, unfamiliar structures individuals tend to leave by the same route they took to enter—perhaps because of their lack of knowledge of their surroundings or perhaps because during extreme situations the brain fixates on the most recent method of entrance. (page 128)

2. Child fire-setters are children ages 2 to 6 who are traditionally curious and set fires in hidden locations, out of sight of adults. Children at this age are not usually able to form the intent of causing damage with fire.

   Fires set by juvenile fire-setters (children aged 7–13) are typically a symptom or indicator of a psychological or emotional problem, often caused by a broken family environment or physical or emotional trauma or abuse. These fires are often set in and around the home or sometimes in an educational setting.

   Fires set by adolescent fire-setters (teens aged 14–16) are usually symptomatic of stress, anxiety, anger, or another psychological or emotional problem. The most usual targets of these fires are schools, churches, vacant buildings, fields, and vacant lots. The fires are frequently associated with disruptive behavior, a broken home life, or a poor social environment. (page 133)

3. Alert words are used to draw the user's attention to the information. Frequently the words "Caution," "Warning," and "Danger" are used as alert words.

   These words are further defined by ANSI Z535.4, *Product Safety Signs and Labels*:

   CAUTION: Indicates a potentially hazardous situation that, if not avoided, may result in minor or moderate injury

   WARNING: Indicates a potentially hazardous situation that, if not avoided, could result in death or serious injury

   DANGER: Indicates an imminently hazardous situation that, if not avoided, will result in death or serious injury. (page 132)

4. When equipment is capable of explosion or starting a fire, the required maintenance becomes critically important. Likewise, the operating procedures for equipment or appliances are designed to ensure safety. If either of these two areas is neglected or improperly performed, the lapse can lead to a fire or explosion. (page 130)

5. Personal interviews can help to establish the following:
   - Prefire conditions
   - Fire and smoke development
   - Fuel packages and their location and orientation
   - Victims' activities before, during, and after discovery of the fire or explosion
   - Actions taken by individuals that resulted in their survival (e.g., escaping or taking refuge)
   - Decisions made by survivors and reasons for those decisions
   - Critical fire events such as flashover, structural failure, window breakage, alarm sounding, first observation of smoke, first observation of flame, fire department arrival, and contact with others in the building (page 133)

Fire Alarms

1. Child fire-setters are children aged 2 to 6 who are traditionally curious and set fires in hidden locations, out of sight of adults. Children at this age are not usually able to form the intent of causing damage with fire.

   Fires set by juvenile fire-setters (children aged 7–13) are typically a symptom or indicator of a psychological or emotional problem, often caused by a broken family environment or physical or emotional trauma or abuse. These fires are often set in and around the home or sometimes in an educational setting.

   Fires set by adolescent fire-setters (teens aged 14–16) are usually symptomatic of stress, anxiety, anger, or another psychological or emotional problem. The most usual targets of these fires are schools, churches, vacant buildings, fields, and vacant lots. The fires are frequently associated with disruptive behavior, a broken home life, or a poor social environment. (page 133)

2. An individual's familiarity with the setting can make escape more likely, although physical limitations and cognitive impairments can minimize the advantages of such familiarity, such as a person's becoming lost in his or her own home. In larger, unfamiliar structures, individuals tend to leave by the same route they took to enter—perhaps because of their lack of knowledge of their surroundings or perhaps because during extreme situations the brain fixates on the most recent method of entrance.

Research has indicated that the degree of familiarity among the individuals in a group also affects response times. If the group is established and its members know one another well, the individuals react and notify each other in a more timely manner than they do if the group is newly formed or its members are unfamiliar with each other. Examples of groups with a high degree of permanence are families, sports teams, choirs, and clubs.

The ability to escape is affected by the identifiability of escape routes, distance to the escape routes, fire conditions (such as smoke, heat, or flames), the presence of dead-end corridors, the presence of obstacles or people blocking the escape path, and the individual's physical disabilities or impairments. (pages 128, 129, and 133)

Chapter 9: Legal Considerations

Matching

1. D (page 142)
2. H (page 141)
3. E (page 145)
4. F (page 145)
5. A (page 141)
6. I (page 145)
7. C (page 141)
8. J (page 146)
9. G (page 143)
10. B (page 141)

Multiple Choice

1. C (page 147)
2. B (page 147)
3. A (page 147)
4. C (page 147)
5. B (page 147)
6. D (page 147)
7. D (page 150)
8. A (page 150)
9. B (page 151)
10. D (page 150)

Fill-in

1. criminal charges (page 139)
2. justified (page 140)
3. consent search (page 140)
4. waiver of rights (page 141)
5. probable cause (page 142)
6. spoliation (page 142)
7. chain of custody (page 145)
8. Testimonial evidence (page 146)

Vocabulary

1. **Arson:** The crime of maliciously and intentionally, or recklessly, starting a fire or causing an explosion. Legal definitions of "arson" are defined by statutes and judicial decisions that vary among jurisdictions. (page 149)
2. **Circumstantial evidence:** Proof of a fact indirectly based on logical inference rather than personal knowledge. (page 150)
3. **Defendant:** The entity against whom a claim is brought in a court. (page 147)
4. **Demonstrative evidence:** Any type of tangible evidence relevant to a case, for example, diagrams and photographs. (page 145)
5. **Deposition:** One type of pretrial oral testimony made under oath. (page 147)
6. **Incendiary fire:** A fire that is intentionally ignited under circumstances in which the person knows that the fire should not be ignited. (page 147)
7. **Tort:** A civil wrong leading to a legal claim for damages. (page 142)

Short Answer

1. Four generally recognized categories of evidence that form the foundation of any civil or criminal fire trial are
    - Real or physical evidence
    - Demonstrative evidence

- Documentary evidence
- Testimonial evidence (page 145)

2. The judge may consider various factors in evaluating an expert's testimony. These factors include, but are not limited to, the following:
   - Whether the theory or technique has been or can be tested and whether the hypothesis underlying the theory or technique can be falsified
   - Whether the theory or technique has been subjected to peer review or publication
   - Whether the theory or technique has a known or potential rate of error
   - The existence and maintenance of standards controlling the technique's operation
   - Whether there has been general acceptance of the theory or technique in the relevant scientific community (page 148)

3. Basic components of an expert's qualifications are as follows:
   - Personal background information (name, age, current place of business or employer, duties in current position)
   - Formal education (schools, colleges, or universities attended, dates attended, diplomas or degrees received, any specialization obtained)
   - Skills and experience (emphasizing those that relate to your qualifications to testify as an expert in fire or explosion cases)
   - Professional licenses or certifications (including the date of any license or certification, the licensing/certifying body, the dates of recertification, and the jurisdictions where the license/certification are recognized—i.e., state, national, international)
   - Distinctions and awards (relating to your work in your field of expertise)
   - Publications (using formal citation format and including the name[s] of any co-authors, title of the work, title of the journal or collection if yours was part of a larger work, name and city of the publisher, publication date)
   - Memberships and offices in professional associations (including committee memberships or chairs and other special appointments. Be careful about listing memberships that were in name only, such as committees to which you were appointed but did not actively participate. Much headway can be made cross-examining you on empty titles.)
   - Professional presentations (speaking engagements at seminars, conferences, and the like)
   - Academic or teaching appointments (page 149)

4. The four examples of circumstantial evidence are as follows:
   - Multiple fires: Two or more separate, nonrelated, simultaneously burning fires
   - Trailers: Deliberately introduced fuel or manipulation of existing fuels to aid in the spread of fire from one area to another
   - Incendiary devices: A mechanism used to initiate an incendiary fire, such as a lit cigarette placed in a book of paper matches and propped over a container of combustible materials, or a Molotov cocktail
   - The presence of ignitable liquids in the area of origin that cannot be explained other than as a device to start or spread the fire (page 150)

5. The following are only some of the potential offenses that may be relevant in a fire investigation:
   - Burning to defraud
   - Insurance fraud
   - Unauthorized burning
   - Manufacture or possession of fire bombs and incendiary devices
   - Wildfire arson
   - Domestic violence
   - Child endangerment/child abuse
   - Homicide and attempted homicide
   - Endangerment/injury to fire fighters
   - Obstruction/interference with fire fighters
   - Disabling of fire suppression/fire alarm systems
   - Reckless endangerment
   - Failure to report a fire
   - Vandalism or malicious mischief
   - Burglary/trespassing (page 150)

Fire Alarms

1. Here is a brief overview of some investigative tasks that may be required:
   - Reviewing NFPA 921 subsection 11.3.5 respecting spoliation of evidence before dealing with products at the fire scene. Note the requirements for notification to interested parties and documenting the product before altering it or conducting any destructive testing.
   - Establishing, as much as possible, the chain of custody of the product as it passed from the manufacturer to the seller and ultimately to the person who owned it at the time of the incident. This means collecting any of the following documents and information that is still available: receipts for the product's purchase, packaging (boxes, warning labels), users' manuals, warranty cards, serial number(s), and information on the make, model, and date of manufacture. In some instances it may even be necessary to trace how the product was transported and warehoused before reaching the consumer.
   - Identifying component parts of the product that may have been defective, resulting in the fire or explosion, and obtaining as many details as possible to determine whether a different company was responsible for manufacturing separate components.
   - Tracing the history of the product, including the use (or misuse) by anyone, including the owner, and the service and maintenance history and records.
   - Retaining one or more experts who can examine the product and/or its components and provide opinions on any defects in the manufacture, warning labels, or design of the product and/or its components. Experts should also be asked to consider whether repairs or modifications were made to the product that contributed to the loss.
   - Obtaining an exemplar of the product so that an undamaged product of the same make, model, and date of manufacture is available for comparison with the fire-damaged product.
   - Maintaining the chain of custody and the integrity of the product after the fire is also critical. It should be documented in place at the scene, using appropriate methods, including photographing and diagramming and perhaps video recording. (page 152)

Chapter 10: Safety

Matching

1. H (page 161)
2. J (page 164)
3. I (page 162)
4. D (page 166)
5. B (page 166)
6. A (page 159)
7. E (page 167)
8. C (page 162)
9. F (page 169)
10. G (page 169)

Multiple Choice

1. A (page 158)
2. B (page 158)
3. D (page 159)
4. C (page 160)
5. C (page 160)
6. A (page 160)
7. B (page 161)
8. D (page 161)
9. C (page 166)
10. A (page 166)

Fill-in

1. shoring (page 166)
2. mitigating the hazard (page 166)
3. incorrectly (page 166)
4. secondary (page 167)
5. contamination (page 169)
6. incident commander (page 169)
7. face-to-face (page 169)
8. safety (page 170)

Vocabulary

1. **Collapse zone:** A zone identified by markers or specialized scene tape that indicates a potential for a structural collapse. (page 159)
2. **Control of Hazardous Energy (Lockout/Tagout) standard:** The federal OSHA regulation that governs work around equipment or wiring where an unexpected energization, startup of machines, or release of stored energy could result in the injury of the investigators. This standard uses a lockout/tagout device that disables the electrical equipment. (page 170)

3. **HAZWOPER (HAZardous Waste OPerations and Emergency Response):** The federal OSHA regulation that governs hazardous materials waste site and response training. Specifics can be found in book 29, standard number 1910.120. (page 170)
4. **Impact load:** A sudden added load to a structure. An example is a fire investigator jumping down onto a floor after peering into the ceiling or over an appliance. (page 166)
5. **Permit-Required Confined Space standard:** The federal OSHA regulation that governs any space that an employee can bodily enter and perform assigned work, with a limited or restricted means of entrance or exit, and that is not designed for continuous employee occupancy. (page 170)

Short Answer

1. The hazards identified on scene may be classified as follows:
   - Physical hazards: Are there many places that an investigator could slip, trip, and fall?
   - Structural hazards: Is the building structurally sound?
   - Electrical hazards: Is the electrical system still connected?
   - Chemical hazards: Is there a material safety data sheet (MSDS), or are there other reference materials available to manage the hazard?
   - Biological hazards: Is there a risk of a hazard caused by bacteria, virus, insects, plants, humans, or animals?
   - Mechanical hazards: Is the equipment still operational or functional? Is the machinery specialized, requiring a technical expert? (page 160)
2. If an occasion arises in which the investigator believes there is a need to enter the building before the fire has been extinguished, the investigator must be equipped with structural firefighting gear and a self-contained breathing apparatus (SCBA). The complete structural firefighting ensemble consists of a helmet, coat, trousers, protective hood, gloves, boots, SCBA, and PASS device. (page 160)
3. Each fire scene is different, and, at times, additional equipment may be needed to provide a safe work environment. All work areas should be well lit to prevent trip-and-fall injuries and to enable the investigator to assess the hazards effectively and complete the investigation. Portable lighting is required at almost every fire scene, but at no time should a generator be brought into an enclosed work area: the buildup of carbon monoxide would quickly become a serious health issue (the action level for carbon monoxide is 25 ppm). Furthermore, when an internal combustion engine is brought into the building, the fire investigator can no longer state that the investigation team did not bring an ignitable liquid into the scene.

   Some investigations require lifelines and fall protection. This equipment should be maintained and stored in accordance with the manufacturer's specifications. The investigator should not depend on any ladders that are present at the fire scene other than those brought by the fire department. Any ladders that were in the fire building may have been exposed to the effects of the fire, or the owner of the building may have abused or improperly maintained the ladders before the fire. If a ladder is required, the investigator should be comfortable with its condition and should ensure that it is the proper height and style for the intended use. (page 163)
4. The investigator who responds to a medical or construction facility can be exposed to radiological materials that are stored at these facilities. In addition, nearly every fatal fire scene contains biological hazards. Rarely does the fire destroy all the bodily fluids of a fire victim. Blood tests show that nearly all fatal fire victims have toxic levels of cyanide in their systems—levels so high that some medical examiners are questioning whether carbon monoxide (generally the most prevalent toxin at a fire scene) or cyanide is killing fire victims. (page 163)
5. An investigator should not enter a burning structure or one that is not completely extinguished without the permission of the incident commander. If such entry is necessary, the investigator should not do so without being accompanied by fire suppression personnel and not unless the investigator is trained to make such entry. (page 165)

Fire Alarms

1. The hazards identified on scene may be classified as follows:
   - Physical hazards: Are there many places that an investigator could slip, trip, and fall?
   - Structural hazards: Is the building structurally sound?
   - Electrical hazards: Is the electrical system still connected?
   - Chemical hazards: Is there a material safety data sheet (MSDS), or are there other reference materials available to manage the hazard?

- Biological hazards: Is there a risk of a hazard caused by bacteria, virus, insects, plants, humans, or animals?
- Mechanical hazards: Is the equipment still operational or functional? Is the machinery specialized, requiring a technical expert?

Although all fire scenes pose some form of hazard to the investigator, the level of risk from each varies. The investigator should assess the hazard present and determine the likelihood that contact would be made with the hazard and, if so, in what form. Knowing this will help to address the measures required to control the hazard. (page 160)

2. When the fire suppression units clear the scene, there must be a face-to-face transfer of command from the incident commander to the fire investigator in charge of the investigation. The transfer of command should identify any areas of concern that can affect the safety of any personnel remaining at the scene. The investigator at that point becomes responsible for the control of the scene and for the safety of everyone at the scene. (page 169)

3. If an occasion arises in which the investigator believes there is a need to enter the building before the fire has been extinguished, the investigator must be equipped with structural firefighting gear and a self-contained breathing apparatus (SCBA).

   The requirements for firefighting PPE are outlined in two standards:

   NFPA 1971, *Standard on Protective Ensembles for Structural Fire Fighting and Proximity Fire Fighting*

   NFPA 1977, *Standard on Protective Clothing and Equipment for Wildland Fire Fighting*

   The requirements for SCBA and PASS devices are outlined in two standards:

   NFPA 1981, *Standard on Open-Circuit Self-Contained Breathing Apparatus (SCBA) for Emergency Services*

   NFPA 1982, *Standard on Personal Alert Safety Systems (PASS)*

   (pages 160 and 161)

Additional Activity

1. The weight of 1 gallon of water is 8.35 lbs.

   A. $(8.35 \times 175) \times 20 = 29{,}255$ lbs
   B. $(8.35 \times 150) \times 15 = 18{,}787.5$ lbs
   C. $(8.35 \times 225) \times 35 = 65{,}756.25$ lbs
   D. $(8.35 \times 250) \times 30 + (8.35 \times 175) \times 40 = 62{,}625 + 58{,}450 = 121{,}075$ lbs

Chapter 11: Sources of Information

Matching

1. H (page 178)
2. I (page 178)
3. J (page 176)
4. B (page 178)
5. C (page 179)
6. G (page 179)
7. A (page 176)
8. D (page 179)
9. F (page 179)
10. E (page 181)

Multiple Choice

1. C (page 179)
2. A (page 179)
3. B (page 179)
4. D (page 179)
5. A (page 179)
6. D (page 179)
7. B (page 179)
8. C (page 179)
9. B (page 181)
10. C (page 181)

Fill-in

1. legal considerations (page 176)
2. Freedom of Information Act (page 176)
3. computers (page 177)
4. as soon as possible (page 177)
5. credibility (page 177)
6. quality; usefulness (page 177)
7. correctly; discussion (page 177)
8. elicit information (page 178)

Chapter 11: Sources of Information

Vocabulary

1. **Confidential communication:** Those statements made under circumstances showing the speaker intended the statements only for the ears of the person addressed. (page 176)
2. **Privileged communication:** Those statements made by certain persons within a protected relationship such as a husband–wife, attorney–client, or priest–penitent. These communications are protected by law from forced disclosure on the witness stand at the option of the witness, spouse, client, or penitent. (page 176)

Short Answer

1. There are numerous groups, services, and organizations that provide services that may be useful during an investigation. Test data, insurance records, and various standards may exist that provide the investigator with valuable resources that governmental agencies may not be able to provide. (page 179)
2. The NFPA develops, creates, and revises various standards and guides that may assist with an investigation. (page 182)
3. National Association of Fire Investigators (NAFI) and International Association of Arson Investigators (IAAI). (page 182)
4. Insurance companies maintain valuable records concerning structures and their owners, which can aid in the detection of fraud and arson (page 182)
5. Many forms of information can be gathered by the investigator, including verbal, written, visual, and electronic information. (page 176)

Fire Alarms

1. **Purpose of Interviews:** Fire investigators may gain crucial information about a fire event during an interview. Witness interviews may reveal possible sources of ignitions, fuel packages and their arrangement within the room of origin, or potential suspects or persons of interest. For example, the fire investigator may wish to interview the last person within a structure or specifically the room of origin to determine any activities or events that may have occurred that could cause the fire. Even witnesses who were not present at the time of the fire may yield valuable information that can assist the investigator. It is important to note that during an interview, the investigator must not only record the information provided but also determine the quality and usefulness of the information.

   Preparation for Interviews: Similar to conducting the scene examination, interviews require a certain amount of skill and training. Investigators must be prepared to conduct an interview and have a thorough understanding of the facts of the incident to that point. These facts allow the investigator to create a plan for conducting the interview, including questions that are crafted to elicit information pertinent to the investigation. Interviews with the incident commander and fire fighters before the scene examination can provide the investigator with information as to the size and extent of the fire on their arrival and unusual circumstances they may have encountered.

   Once a witness has been identified, he or she should be interviewed as soon as possible to ensure that the facts are correctly remembered by the witness and not clouded by discussion with other witnesses. These interviews often occur at the scene; however, other locations may be more suitable, such as a designated interview room. If needed, follow-up interviews can be conducted at a later time that may be more practical. Interviews are as dynamic as the scene itself and require the investigator to remain flexible and adaptive to each witness. The investigator should inform the witness who he or she is and provide appropriate credentials when requested. Likewise, the investigator should also establish the identity of all persons interviewed as well as the information in Table 11-1 in the textbook. This information may prove useful at a later date if you need to conduct a follow-up interview or locate the witness for the purpose of testifying in court. It may be several months or years after the fire that an arrest or trial occurs. Key witnesses may have changed jobs or relocated or may become reluctant to participate in that time.

   Questions should be meaningful and designed to elicit information from the witness. Open-ended questions work best with most witnesses because the witnesses are allowed to tell their answer almost like a story. Closed-ended questions generally produce only one- and two-word answers that lack meaning and explanation. For example, the investigator may wish to know what the witness saw. A closed-ended question may be "Did you see the fire?," whereas an open-ended question might be "What exactly did you see?" Both question types are useful when appropriate. Open-ended questions work best when attempting to obtain information during initial interviews and in follow-ups when new information is requested. Closed-ended questions are useful for very distinct responses where "yes" or "no" will suffice. (page 177)

Answer Key

2.

Municipal clerk	Maintains all records related to municipal licensing and municipal operations.
Municipal assessor	Maintains all public records related to real estate, including plot plans, maps, and taxable real property.
Municipal treasurer	Can provide public records related to names and addresses of property owners, legal descriptions of property, and the amount of paid or owed taxes on a property.
Municipal Street Department	Maintains records and maps of municipal conduits, drains, sewers, street addresses, and all old and current street names, including alleys and right of ways.
Municipal Building Department	Has records related to building, electrical, and plumbing permits and archive building blueprints and files.
Municipal Health Department	Maintains records of births and deaths and investigations related to health hazards.
Municipal Board of Education	Contains records related to the school system and may assist with identifying and locating school-age offenders.
Municipal Police Department	Can provide records related to local criminal investigations
Municipal Fire Department	Maintains records related to fire and EMS incidents and life-safety inspections.
Other municipal agencies	Varies from government entities but may include public works, parks and recreation, and water distribution.
County recorder	Responsible for recording legal documents that determine ownership of real property and maintains files of birth, death, and marriage records as well as bankruptcy documents.
County clerk	Maintains public records related to civil litigation, probate records, and other documents related to county business.
County assessor	Maintains records related to property and plats, including property owners, addresses, and taxable value.
County treasurer	Can provide information related to property owners, tax mailing addresses, legal descriptions, and the amount of either owed or paid taxes on property, and also maintains all county financial records.
County coroner/medical examiner	Can provide information related to the identification of victims, manner and cause of death, as well as any items found either near or on the victim.
County Sheriff's Department	Can provide both investigative and technical support for county criminal investigations and provides polygraph services, evidence collection, and retention.
Other county agencies	Various departments include parks and recreation, conservancy districts, Homeland Security, and Emergency Management.
Secretary of State	Maintains records related to charters and annual reports of corporations, charters of villages and cities, and trade name and trademark registrations
State treasurer	Maintains public records related to state financial business.
State Department of Vital Statistics	Maintains records for births, deaths, and marriages.
State Department of Revenue	May assist with locating tax records of individuals or corporations, both past and present, as well as locating individuals through child support records.
State Department of Regulation	Source of information such as professional licenses, results of licensing exams, and regulated businesses.
State Department of Transportation	May provide information about highway construction and improvements, motor vehicle accident investigations, and vehicle registration and operator testing and regulations.

State Department of Natural Resources	Responsible for conservation and protection of water lands, rivers, lakes, and forest areas and may provide records related to hunting and fishing licenses, waste disposal regulations, and cooperation with the Environmental Protection Agency (EPA).
State Insurance Commissioner's Office	Can provide assistance related to licensed insurance companies, insurance agents (both past and present), and computer complaints.
State Police	May provide information related to state criminal investigations. Some state police agencies may also conduct fire investigations or provide assistance to local agencies. Additionally, they may also operate their own forensic laboratory.
State Fire Marshal's Office	Can provide information in regard to fire incidents within the state, building inspection records, fireworks and pyrotechnics, and boiler inspections. Most state fire marshals also have fire investigators who may provide assistance with conducting origin and cause investigations.
Other state agencies	May include the Department of Motor Vehicles, Department of Homeland Security, Liquor Enforcement.
Department of Agriculture	Maintains records related to food stamps, meat inspections, and dairy products and oversees the U.S. Forestry Service.
Department of Commerce	Maintains records related (but not limited) to highway projects and names and addresses of ships fishing in local waters, trade lists, and patents
Department of Defense	May provide public records related to the five military branches. Also, all branches maintain their own investigation units.
Department of Health and Human Services	Maintains records related to social security and the Food and Drug Administration and maintains an investigation unit.
Department of Housing and Urban Development	Maintains records related to public housing and federal assistance.
Department of the Interior	Maintains records related to fish and game activities and Indian Affairs. The National Park Service is maintained within this department.
Department of Labor	Can provide information related to labor and management, including overtime, pay, and combat age discrimination.
Department of State	Can assist with investigations by locating visas of foreign nationals and companies that operate within the country and abroad.
Department of Transportation	Can provide records related to vehicle transportation and hazardous materials.
Department of the Treasury	May provide technical information through the Bureau of Alcohol, Tobacco, Firearms, and Explosives as well as license holders, manufacturers, and importers of firearms. The U.S. Customs service maintains records related to importers and exporters, customhouse brokers, and truckers, as well as licensing vessels not licensed by the U.S. Coast Guard. The Internal Revenue Service may assist with matters related to federal income tax, and the U.S. Secret Service maintains public records related to counterfeiting and forgery of U.S. currency and conducts investigations related to threats against all current and former presidents and their families as well as foreign heads of state.
Department of Justice	Assists with records related to antitrust and civil rights violations. This department includes the Civil Rights Division, the Criminal Division, the Drug Enforcement Administration, the Federal Bureau of Investigation, and the Immigration and Naturalization Service.

Answer Key

U.S. Postal Service	Postal inspectors may provide information that has been routed through the mail system.
Department of Energy	Provides information related to the nation's energy policies and programs.
United States Fire Administration	Oversees the National Fire Incident Reporting System (NFIRS) and the Arson Information Management.
National Oceanic and Atmospheric Administration	Maintains records of past and present weather data.
Other federal agencies	Various other federal agencies may provide the investigator with public information useful for their investigation.
National Fire Protection Association (NFPA)	Develops, creates, and revises various standards and guides that may assist with an investigation.
Society of Fire Protection Engineers (SFPE)	Works to advance fire protection engineering, including publishing various documents that may prove useful to the investigator.
American Society for Testing and Materials (ASTM)	Develops voluntary consensus standards that may be used by architects during the design phase as well as procedures for fire tests.
National Association of Fire Investigators (NAFI)	Provides training related to fire investigation topics and implemented the National Certification Board.
International Association of Arson Investigators (IAAI)	Provides training programs for investigators including the CFITrainer.net online for fire investigators.
	Provides professional credentials for fire investigators and works to control arson and other related crimes.
Regional fire investigations organizations	May exist at either the state or town level and provide contacts that can be consulted
Real estate industry	Maintains valuable records concerning structures and their owners, which can aid in the detection of fraud and arson.
Abstract and title companies	Maintain records related to former and current property owners, as well as escrow account maps and tract books.
Financial institutions	Maintain various records of both individuals' and businesses' financial records, as well as information of loan companies, brokers, and transfer agents.
Insurance industry	Maintains valuable records concerning structures and their owners, which can aid in the detection of fraud and arson.
Educational institutions	Can provide information related to a person's background and personal interests.
Utility companies	Maintain databases of customers, as well as documented problems on the status of their distribution equipment.
Trade organizations	Act as a clearinghouse of information specific to their discipline and usually create and publish trade magazines.
Local television stations	Often provide investigators with copies of videotape related to an incident.
Lightning detection networks	Lightning and weather data can play an important part in causal analysis of a fire scene.
Other private sources	Numerous other private sources exist that the investigator may find helpful depending on the specific needs of the investigation. (pages 178–182)

Chapter 12: Planning and Preplanning the Investigation

Matching

1. D (page 189)
2. E (page 188)
3. F (page 189)
4. C (page 190)
5. A (page 188)
6. B (page 189)

Multiple Choice

1. C (page 193)
2. A (page 193)
3. D (page 193)
4. B (page 193)
5. A (page 193)
6. C (page 193)
7. B (page 193)
8. D (page 193)
9. A (page 193)
10. C (page 193)

Fill-in

1. diverse skills (page 188)
2. safety issues (page 188)
3. resources (page 188)
4. team concept (page 189)
5. unique (page 189)
6. personnel or resources (page 189)
7. competencies (page 190)
8. Networking (page 192)

Vocabulary

1. **Preinvestigation team meeting:** A meeting that takes place before the on-scene investigation. The team leader or investigator addresses questions of jurisdictional boundaries and assigns specific responsibilities to the team members. Personnel are advised of the condition of the scene and the safety precautions required. (page 190)

Short Answer

1. It is important to use the team concept whenever possible. It is recognized that the investigator may be required to perform all functions of an investigation and must preplan for that type of investigation as well. A fire scene investigation includes photography, sketching, evidence collection, witness interviews, and other varied tasks that require diverse skills. Through use of the team concept, the investigator in charge can delegate these functions to the individuals best qualified to perform them and thus ensure a thorough and professional investigation. (page 188)
2. During the initial assignment and response, basic incident information obtained should include the following:
    - The location of the incident
    - The date and time of the incident
    - The weather conditions at the time of the incident
    - The size and complexity of the incident
    - The type and use of the structure involved in the incident
    - The nature and extent of the damage
    - The security of the scene
    - The purpose of the investigation
    (page 189)
3. Security will almost always be an issue, both before and during the investigation. Potential safety issues should never be overlooked. Fire scenes provide for hazards that are often unique and must be preplanned. Some of the more common safety issues include overhead hazards, foot hazards, slips and falls, and electrical issues. However, fire investigations are often in areas of other hazards, such as the hazardous chemicals that may have been present before the fire and those that have developed during the fire. These hazards may require that special personal protective equipment (PPE) be used. Investigators operating alone should also understand these issues and prepare for them. (page 189)
4. Six basic functions are commonly performed in each investigation:
    - Leadership and coordination
    - Safety assessment

- Photography, note taking, mapping, and diagramming
- Interviewing witnesses
- Searching the scene
- Evidence collection and preservation (page 190)

5. A number of pieces of protective equipment are recommended to be worn by fire investigators:
   - Eye protection
   - Flashlight
   - Gloves
   - Helmet or hard hat
   - Respiratory protection (the type depends on the exposure)
   - Safety boots or shoes
   - Turnout gear or coveralls (page 192)

Fire Alarms

1. Preplanning for expertise of various fields will provide resources when needed to respond to a particular type of incident. This may include engineers or scientists of various backgrounds, for example, electrical wiring systems, electrical equipment, heating equipment, forensic debris analysis, forensic body analysis, fire suppression and detection equipment, fire dynamics, and fire spread. (page 188)

2. Fire scene investigators must recognize the limitations of their expertise and preplan for resources to fill those limitations. Limitation is not a problem as long as it is recognized and resources are provided to supplement the limitation. Problems are likely to occur if an investigator goes beyond his or her expertise. For example, most investigators do not have a background or training in evaluating failure modes in the electrical systems of mechanical equipment. Identifying the requirements and the type of investigation you may be assigned requires the investigator to identify his or her own limitations.

   Networking with other investigators or associations is of particular benefit to identifying resources in meeting these needs and limitations. These investigation networks should be noted during preplanning in the event that the investigator finds the need for additional resources or expertise that has not been previously developed. (page 192)

Chapter 13: Documentation of the Investigation

Matching

1. G (page 199)
2. C (page 199)
3. J (page 198)
4. H (page 200)
5. F (page 200)
6. B (page 200)
7. I (page 199)
8. A (page 199)
9. E (page 200)
10. D (page 200)

Multiple Choice

1. C (page 202)
2. B (page 202)
3. A (page 202)
4. C (page 200)
5. B (page 199)
6. D (page 200)
7. A (page 200)
8. D (page 202)
9. B (page 205)
10. C (page 205)

Fill-in

1. accepted practices (page 200)
2. sequential views (page 200)
3. diagram of the site (page 200)
4. legend (page 205)
5. specification sheet (page 205)
6. note taking (page 207)
7. report (page 207)
8. Computer presentations (page 204)

Vocabulary

1. **Bracketing:** Taking a series of photographs with sequentially adjusted exposures. (page 199)
2. **Photo painting:** A technique used when the photo will cover a large scene. It can be accomplished by placing the camera in a fixed position with the shutter locked open. A flash unit can be fired from multiple angles to illuminate multiple subjects or large areas from all angles. (page 200)
3. **Ring flash:** Reduces glare and gives adequate lighting for the subject matter. (page 200)
4. **Sequential photography:** Shows the relationship of a small subject to its relative position in a known area. The small subject is first photographed from a distant position, where it is shown in context with its surroundings. Additional photographs are then taken increasingly closer until the subject is the focus of the entire frame. (page 200)

Short Answer

1. Photographs are an integral part of the examination and should reflect the condition of the scene as seen by the investigator. The photographs should be taken in a predetermined manner and in accordance with accepted practices in the fire and explosion investigation field—for example, by illustrating the philosophy of examination from areas of least damage to areas of most damage.

   An effective general technique at a fire scene is to begin film documentation on the exterior perimeter, progressing to the interior, and from the least damaged areas to the most damaged areas in a sequential manner. Critical evidence should be documented by photographing the subject from different angles, including downward from a ladder. (page 200)

2. When recording the scene the investigator should annotate a diagram of the site, identifying the point from which each photograph was taken, the direction of the photograph, the placement of the item, and the photo number. (page 200)

3. It is important to document as much of the scene as possible. Some suggested activities to record are conditions on arrival, suppression, overhaul, observers, and origin and cause determination. The progression of the fire, its colors, its reaction to suppression activities, and the overhaul procedures used are all important in helping the fire investigator determine the origin and cause. Photographs can also document the extent of damage to the victims or structure.

   Photographs of the crowd observing the fire scene activity can help the investigator identify individuals who may have knowledge beneficial to the investigation. These photographs can also help the investigator identify individuals who are seen at multiple fires or are known by the law enforcement or fire department community. (page 202)

4. Exterior photographs are important and can be used to establish the location of the fire scene. Exterior photographs should include street signs, house numbers or other identifiable landmarks that are likely to remain for some time, surrounding locations of the fire scene, and all angular views of the exterior of the fire scene. Exterior photos could also be taken of the address numbers of affected buildings. This may be useful for documentation and search warrant purposes. (page 202)

5. Diagrams are formal drawings that are completed after the investigation. Sketches are generally freehand diagrams or diagrams drawn with minimal tools that are completed at the scene. The differences among the types of drawings relate to the amount of detail that is incorporated, the type of construction of the structure, features of the structure, equipment, and other factors that are important as to the origin, cause, and spread of the fire. Every investigation should include fire scene sketches. This is especially important for investigations likely to be involved in criminal or civil litigation. (page 204)

Fire Alarms

1. It is important to document as much of the scene as possible. Some suggested activities to record are conditions on arrival, suppression, overhaul, observers, and origin and cause determination. The progression of the fire, its colors, its reaction to suppression activities, and the overhaul procedures used are all important in helping the fire investigator determine the origin and cause. Photographs can also document the extent of damage to the victims or structure.

Photographs of the crowd observing the fire scene activity can help the investigator identify individuals who may have knowledge beneficial to the investigation. These photographs can also help the investigator identify individuals who are seen at multiple fires or are known by the law enforcement or fire department community.

All fire investigators should routinely determine whether initial witnesses to the fire such as the 911 reporting party documented what they saw with a camera phone or digital camera. Many times media news outlets have staff or stringers respond to large fires. These professionals sometimes arrive on scene before apparatus and often have superior quality photographic equipment. Reporters may be busy interviewing victims or witnesses before your arrival and should always be contacted.

Photographs of the suppression activities can help the fire investigator understand why the fire reacted in a particular manner. Also, when documenting suppression activities, the location of hydrants, engine companies, apparatus, and hose placement should be documented. (page 202)

2. Photographs should also be taken of occupant or victim actions and locations at the scene. These pictures should document any survivor's or victim's location at the time of the fire, any indicators of actions taken by them during the fire, and any result such as serious injury or death. If the scene includes a fatality, the position of the victim should be thoroughly recorded. If the full condition of the victim cannot be documented at the scene because of lighting, scene hazards, or other obstacles, additional photos should be taken at the medical examiner's facility.

When a witness or victim reports that he or she saw a potentially significant event, the photographer should attempt to photograph the view from the position of the witness.

Aerial photographs can help the investigator clarify scene arrangement and large evidence items such as a vehicle or body. (page 204)

3. Depending on the complexity of the investigation, there are many elements that should be included on sketches and diagrams:
   - General information: The investigator should identify the name of the person who created the diagram, the diagram title, and date of preparation.
   - Identification of compass orientation: Most often the sketch will note north at the top of the page.
   - Scale: The investigator should draw to scale and should indicate on the drawing whether it is "not to exact scale" or "approximate scale."
   - Symbols: The investigator must be consistent with the use of symbols, not using the same symbol for multiple purposes. It is recommended that the investigator use symbols used in engineering or architecture. Fire protection symbols can be found in NFPA 170, *Standard for Fire Safety and Emergency Symbols*.
   - Legend: The fire investigator should create a legend for any drawing, indicating what the referenced symbols represent. (page 205)

Chapter 14: Physical Evidence

Matching

1. H (page 212)
2. D (page 214)
3. J (page 212)
4. F (page 214)
5. B (page 214)
6. A (page 212)
7. I (page 218)
8. E (page 219)
9. C (page 216)
10. G (page 222)

Multiple Choice

1. B (page 212)
2. B (page 212)
3. C (page 214)
4. D (page 214)
5. A (page 215)
6. C (page 215)
7. A (page 215)
8. B (page 217)
9. D (page 217)
10. C (page 217)

Fill-in

1. firefighting (page 212)
2. judge or jury (page 214)
3. area of origin (page 214)
4. legal sanctions (page 214)
5. potential evidence (page 215)
6. fixed in a diagram (page 216)
7. clean collection tools (page 217)
8. Traditional forensic physical evidence (page 218)

Vocabulary

1. **Accelerant:** Any fuel or oxidizer, often an ignitable liquid, used to initiate a fire or increase the rate of growth or speed the spread of fire. (page 218)
2. **Cross-contamination:** The unintentional transfer of a substance from one fire scene or location contaminated with a residue to an evidence collection site. (page 216)
3. **Demonstrative evidence:** Photographs, maps, X-rays, visible tests, and demonstrations. (page 214)
4. **Headspace:** The zone inside a sealed evidence can between the top of fire debris and the bottom of the lid. (page 219)
5. **Traditional forensic physical evidence:** Includes, but is not limited to, finger and palm prints, bodily fluids such as blood and saliva, hair and fibers, footwear impressions, tool marks, soils and sand, woods and sawdust, glass, paint, metals, handwriting, questioned documents, and general types of trace evidence. (page 218)

Short Answer

1. Key properties of common ignitable liquids include the following:
   - Liquids flow downgrade and tend to form pools or puddles in low areas.
   - Almost all hydrocarbon liquids are lighter than water, are immiscible, and may display "rainbow" coloration (sheen floating on water). Certain other common ignitable liquids (e.g., alcohol and acetone) are water soluble.
   - Almost all commonly used ignitable liquid accelerants tend to form flammable or explosive vapors at room temperature.
   - The vapors of most commonly used ignitable liquid accelerants are heavier than air and tend to flow downward into stairwells, cellars, drains, pipe chases, elevator shafts, and so on.
   - Many ignitable liquids used as fire accelerants are readily absorbed by structural materials and natural or synthetic substances.
   - Many ignitable liquids are powerful solvents, which tend to dissolve or stain many floor surfaces, finishes, and adhesives.
   - Common ignitable liquids used as fire accelerants do not ignite spontaneously.
   - Ignition of a given ignitable liquid vapor requires that the vapor be within its flammable or explosive range at the point where it encounters an ignition source at or above its ignition temperature.
   - When an ignitable liquid is poured onto a floor and ignited, two major things occur:
     - Many types of synthetic surfaces (e.g., vinyl) or surface treatments mollify (soften) beneath the liquid.
     - At the edges of the pool, burning vapors adjacent to the liquid edge cause many floor surfaces, such as wood, to char, whereas certain others, such as vinyl, melt and then char. As the liquid pool boils off, its edge recedes. Floor surface charring (or melting and charring) follows the receding liquid edge. The floor area under the ignitable liquid is protected from the effects of burning until the liquid boils off that section.
   - Experiments indicate that the greatest temperatures in an ignitable liquid-accelerant fire occur above the center of the burning liquid pool. Scientific experiments have shown that maximum concentrations of ignitable liquid residues are more often found at the edges of the burn pattern and minimum concentrations toward the center. Some arson investigators believe this is controversial and so take samples from both the edges and the center.
   - Ignitable liquids with high vapor pressure, such as alcohol or acetone, tend to "flash and scorch" a surface, whereas ignitable liquids with higher boiling components, such as kerosene or turpentine, tend to "wick, melt, and burn," leaving stronger patterns. The amount of ventilation available to the fire is a factor in burn pattern appearance. (page 218)
2. Six desirable ignitable liquid collection areas are as follows:
   - Lowest areas and insulated areas within the fire damage pour pattern
   - Samples taken from porous plastic or synthetic fibers
   - Cloth, paper, and cardboard in direct contact with the pattern
   - Inside seams, tears, and cracks
   - Edges of burn patterns
   - Floor drains and bases of load-bearing columns or walls (page 219)

3. In cases in which a container is suspected of containing an ignitable liquid used as an arson accelerant or in which such a liquid was a factor in an accidental case, there is a series of recommendations for evidence recovery:
   - Find out whether investigators recognize the odor of the liquid to allow later testimony about odor recognition.
   - Collect a sample of the liquid into a sterile glass container with a hard plastic cap by pouring or drawing the liquid into a sterile pipette or eyedropper. It is good practice to place a piece of aluminum foil on top of the container before screwing down the cap.
   - Remove the container from the fire scene. Pour a small amount onto a safe surface, and attempt to ignite it to allow later testimony about ignition properties.
   (page 219)
4. After a fire has been brought under control, interior crews may begin to discover potentially important physical evidence. There are a number of effective ways to protect physical evidence on a fire ground. The best way is to post a fire fighter at the entrance to the area of fire origin or near a critical piece of evidence to restrict access, create a fire watch, and provide additional suppression as required. Trained investigators should be called to the fire scene as early as possible in this process.

   Potential physical evidence, such as a gasoline container, a gun, a tool that may have been used in a burglary, an appliance that shows evidence of electrical faulting, or similar item, can often be "taped off" or identified by an evidence cone. The incident commander (IC) should always be notified when potential evidence is discovered. (page 214)
5. When marking an item of physical evidence, tag, or package for identification, the following data should be included:
   - Date and time collected
   - Case number
   - Location
   - Brief description of the evidence
   - Where and at what time the item was discovered
   - Name of the investigator(s) collecting the evidence (page 221)

Fire Alarms

1. Common physical evidence at a fire scene may include traces of ignitable liquid in flooring, a tool mark that is at a point of forcible entry or that indicates adjustment of a critical valve, a faulted electrical circuit, fingerprints, blood, or other physical item or mark that helps the investigator establish fact. (page 212)
2. The first stage of preservation of potential physical evidence on a fire ground begins with the firefighting or suppression operation. Although fire fighters are not responsible for determining the origin or cause of fire, they play an integral part in the investigation. Because the specific cause of a fire may not be known until after a full scene examination is conducted, a professional approach is to consider the entire fire scene as physical evidence to be protected to the extent possible by controlled firefighting and conservative overhaul operations, especially in the area of fire origin. Fire fighters are trained to look for indicators of incendiarism, including multiple fires, incendiary devices, or ignitable liquids.

   In initial fire ground operations, the security of each door and window is an issue. During the size-up, well-trained fire fighters should always take note of existing conditions during their first contact with the scene, including audible alarms, open or unlocked doors or windows, odors, and objects that do not seem to belong where they are located. Once life and safety issues are controlled, the IC should ensure that the fire scene is protected from any further destruction. (pages 214–215)
3. Close coordination of fire suppression and fire investigation in a community can minimize the "time out of service" issue for fire suppression crews while maximizing the fire prevention, code enforcement, and law enforcement benefits of an effective fire investigation program. The solution to many of these traditional problems is to call experienced fire investigators to the scene as early in the firefighting process as possible. Inadvertent damage to physical evidence and loss of potential witnesses can be avoided by early intervention.

   Before any evidence is removed from the area of fire origin, it needs to be analyzed and photographically documented and fixed in a diagram that indicates its location and position before it is collected. Field note taking should document the condition of the evidence when it is discovered and should list the other people who are present.

Skilled investigators, prosecutors, and medical examiners caution against moving physical evidence, such as containers of a suspected accelerant or dead bodies, from a fire scene before they can be carefully examined and their location documented.

Likewise, in cases in which there is evidence of forcible entry, pieces of an explosive device, or some other potential item of physical evidence, the general rule is that the evidence should be left undisturbed and firefighting operations, whenever possible, should be directed to a secondary position away from potential evidence sites. (pages 215–216)

Additional Activity

1.
Crime Scene Search Evidence Report

Name of subject John Doe
Offense Arson
Date of incident 4 July 2010   Time 03:30 a.m. a.m. p.m.
Search officer Fire Marshall Jones
Evidence description 5 gallon red steel gasoline can with steel flexible spout
Location 1234 South Main Street, Anytown, U.S.A.

Chain of Possession

Received from Lieutenant John Smith, E1, Anytown FD
By Chief James Brown, Anytown FD
Date 4 July 2010   Time 03:30 a.m. a.m. p.m.

Received from Chief James Brown, Anytown FD
By Fire Marshall Jones
Date 4 July 2010   Time 03:50 a.m. a.m. p.m.

Received from Fire Marshall Jones
By ABC Forensic Laboratories
Date 6 July 2010   Time 10:00 a.m. a.m. p.m.

Received from ABC Forensic Laboratories
By Fire Marshall Jones
Date 9 July 2010   Time 1:30 p.m. a.m. p.m.

Chapter 15: Origin Determination

Matching

1. A (page 228)
2. J (page 229)
3. H (page 238)
4. F (page 235)
5. G (page 229)
6. C (page 228)
7. D (page 240)
8. I (page 236)
9. E (page 240)
10. B (page 230)

Multiple Choice

1. D (page 228)
2. B (page 229)
3. A (page 229)
4. C (page 230)
5. C (page 231)
6. B (page 233)
7. D (page 233)
8. A (page 235)
9. D (page 235)
10. C (page 236)

Fill-in

1. does not (page 229)
2. removal of debris (page 232)
3. outside (page 232)
4. fire scene reconstruction (page 233)
5. inaccurate (page 233)
6. intensity patterns (page 236)
7. heat and flame vector (page 235)
8. physical evidence (page 238)

Vocabulary

1. **Arc survey diagrams:** Locations of electrical arcing are identified and plotted on a diagram of the affected area of the structure. The spatial relationship of the arc sites can create a pattern and help establish the sequence of damage. This analysis can be used on building circuits and electrical devices within a compartment to help identify or narrow the area of origin. (page 233)
2. **Area of origin:** The room, building, or general area in which the point of origin is located. (page 228)
3. **Depth of char surveys:** Measurements of the relative depth of char on identical fuels are plotted on a detailed scene diagram to determine locations within a structure that were exposed longest to a heat source. (page 233)
4. **Depth of calcination surveys:** Measurements of the relative depth of calcination (observable physical changes in gypsum wallboard) are plotted on a detailed scene diagram to determine locations within a structure that were exposed longest to a heat source. (page 233)
5. **Fire spread analysis:** The process of identifying fire patterns relating to the movement of fire from one place to another and the sequence in which the patterns were produced to trace the fire back to an origin. (page 229)
6. **Isochar:** A line on a diagram connecting equal points of char depth. (page 235)
7. **Point of origin:** The exact physical location where a heat source and a fuel come into contact with each other and a fire begins. (page 228)
8. **Safety assessment:** Inspection to determine whether the scene is safe to enter and the steps necessary to render the scene safe. (page 230)
9. **Sequential pattern analysis:** Application of principles of fire science to the analysis of fire pattern data (including fuel packages and geometry, compartment geometry, ventilation, fire suppression operations, witness information, etc.) to determine the origin, growth, and spread of a fire. (page 229)
10. **Total burn:** A fire scene where a fire continued to burn until most combustibles were consumed and the fire self-extinguished because of a lack of fuel or was extinguished when the fuel load was reduced by burning and there was sufficient suppression agent application to extinguish the fire. (page 238)

Short Answer

1. Relevant information relating to origin can be obtained from one or more of the following sources: fire patterns, arc mapping, heat and flame, vector analysis, depth of char/calcination survey, application of fundamental principles of fire dynamics, and witness information. (page 228)
2. The process for data collection, as it relates to origin determination, includes the initial scene assessment, safety assessment, excavation and reconstruction, and collection of data from nonscene sources. (page 229)
3. Documenting various exterior building features include, but are not limited to, the following:
   - Prefire conditions: The state of repair of the building and its components, the conditions of structural elements, the condition of the structure's fire suppression, and detection systems
   - Utilities: Type, location, and meter readings
   - Doors, windows, and other openings: Their location, condition (open, closed, or broken, and for how long prior to the fire), and security mechanisms
   - Explosions: The presence or absence of evidence that explosions may have affected the exterior components
   - Fire damage: Overall damage, damage to natural openings (windows, doors), damage resulting in unnatural openings (holes made by the fire or explosion, holes made during suppression efforts) (page 230)

4. All interior rooms and other relevant areas should be inspected to identify areas requiring detailed inspection. Interior damage should be compared with the location and type of damage on the exterior. Among the things to examine are the following:
   - Prefire conditions of the structure
   - The contents of the structure
   - Storage of contents
   - Housekeeping
   - Maintenance
   - Evidence of explosion damage
   - Areas of fire damage
   - Building systems (HVAC, electrical, fuel gas, fire protection, alarms, and security)
   - The composition of the surface coverings of walls and floors
   - Position of windows, doors, and other openings (ventilation aspects)
   - Indicators of smoke and heat movement (i.e., fire patterns) on various surfaces
   - Relative extent of damage (severe, moderate, minor, none) in each area (page 231)
5. Adequate removal of debris is essential in leading the investigator to the correct analysis. Fire suppression crews should attempt to minimize contents and debris removal during overhaul before the start of the investigation. If this is not possible, the investigator should document the conditions before debris removal. This documentation should include photographs and notes.

   The debris should be moved only once. It should be removed in a systematic fashion, and the process should be fully documented, including the location, condition, and orientation of any contents that are uncovered. The investigation team should work together, deciding what is important and what is not before beginning the process. Hand tools rather than heavy equipment should be used to remove debris, when possible. (page 232)

Fire Alarms

1. The determination of origin begins with the initial scene assessment. The initial scene assessment provides an overall look at the structure and surrounding area. In this step data collection for the determination of origin begins. The purpose of this assessment is to determine the scope of the investigation, for example, equipment, manpower, safety, and security requirements. Furthermore, it helps to identify areas that require detailed inspection.

   The first order of business in the initial scene assessment is the safety assessment. During the safety assessment of the scene, the investigator will need to assess each of the hazards. The overall goal is to determine whether it is safe to enter the scene, what steps are required to make the scene safe to enter, and what personal protective equipment (PPE) will be required when entry is made. Only after appropriate safety measures have been implemented should the initial scene assessment continue.

   Beginning with the initial scene assessment, it may be beneficial to establish building identifiers and to use them throughout the investigation. These identifiers provide consistency in various aspects of an investigation—such as in the description of photographs, diagram preparation, report writing, and testimony. Building identifiers or a building marking system can include any of the following:
   - Points on a compass (north side of the building, etc.)
   - Front, back, left side, right side
   - Similar methods such as those used in the incident command (side A, side B, etc.) or for urban search and rescue

   The initial scene assessment should include examination of the site and of surrounding areas. This involves looking for fire patterns away from the core of the scene. Important witnesses can be identified, among them the reporting party and neighbors. It is important for the investigator to document findings during this phase.

   As part of the initial scene assessment the investigator should walk around the entire exterior to evaluate the possibility of extension from an outside fire, to examine the building construction details and materials, to determine the occupancy or use of the building, and to identify areas that may require further study. During this and all subsequent phases the investigator should constantly be evaluating the safety of the building and identifying any conditions that need to be corrected. These conditions include securing utilities, shoring unstable areas, removing water, cordoning off unsafe areas, and noting inhalation dangers.

The weather at the time of the fire is an important component of the initial scene assessment. Wind can play a role in the development of the fire and the patterns left behind. Weather information can be obtained from a number of sources, including the local airport, local media outlets, the National Weather Service, and Internet sources. Another source would be the weather observations of witnesses and fire fighters obtained through interviews. It is important to consider the distance between the point at which the weather readings were taken and the fire scene. Weather conditions can vary dramatically from one location to another. The investigator should corroborate official reports with observations from local witnesses. (page 230)

2. Potential ignition sources, initial fuels, or other important evidence can be identified during excavation of the scene. Care should be taken to limit damage to and prevent contamination of any such items. Simple decontamination and processing techniques can prevent these problems.

Washing floors after debris removal may help reveal fire patterns. Caution should be exercised when using high-pressure streams, as they can damage evidence. Washing should be done only after adequate debris removal, after samples have been collected, and after proper scene documentation has been completed.

Both public- and private-sector investigators should recognize that civil litigation may result from fire incidents. Parties that may be interested in such litigation should be identified, placed on notice, and invited to participate in the ongoing inspection. It is important that potentially interested parties be provided the opportunity to view the scene and evidentiary items in place where possible. It is understood that different requirements for criminal investigations exist. (pages 232–233)

Chapter 16: Fire Cause Determination

Matching

1. C (page 245)
2. D (page 246)
3. B (page 246)
4. A (page 246)
5. F (page 245)
6. E (page 246)

Multiple Choice

1. A (page 245)
2. D (page 245)
3. C (page 245)
4. B (page 245)
5. A (page 245)
6. C (page 245)
7. A (page 246)
8. D (page 246)
9. B (page 246)
10. C (page 246)

Fill-in

1. ignition sources (page 244)
2. hypothesis (page 244)
3. physical evidence (page 245)
4. not sufficient (page 245)
5. remote (page 246)
6. ignition temperature (page 245)
7. thermal inertia (page 245)
8. configuration (page 245)
9. migrate (page 246)
10. possible; probable (page 246)

Vocabulary

1. **Fuel analysis:** Identification of the first fuel ignited. (page 246)
2. **Ignition sequence:** The sequence of events and circumstances that allow the initial fuel and the ignition source to come together and result in a fire. (page 248)
3. **Thermal inertia:** The product of thermal conductivity, density, and specific heat (or heat capacity). (page 245)

Short Answer

1. Any identified ignition source must have sufficient temperature and energy and must have the ability to raise the fuel to its ignition temperature. This process is identified with three steps:
   - Generation: The competent ignition source must generate energy in the form of heat and be able to transfer sufficient energy to a fuel package to reach its ignition temperature.

- Transmission: Whether by conduction, convection, radiation, or direct flame contact, the heat source must be able to transmit enough energy to the fuel to reach its ignition temperature.
- Heating: Thermal inertia—the product of thermal conductivity, density, and specific heat—will influence the reaction of a fuel package to exposed heat.(page 245)

2. Essential data to be collected to determine cause include identification of fuels, potential ignition sources, and unusual oxidants in the area of origin. (page 246)
3. Factors to be considered in developing an ignition sequence include the following:
   - How and when the initial fuel came to be present at the origin
   - How and when the oxidant came to be present (if unusual oxidant is involved)
   - How and when the competent ignition source came to be present
   - How and when the competent ignition source transferred heat to the initial fuel
   - What acts or omissions, and in what sequence, brought together the fuel, oxidizer, and ignition source
   - How the initial fuel ignited subsequent fuels to result in fire spread (page 248)
4. The investigator may establish one of two levels of certainty based on his or her confidence in the data collected:
   - Probable: A level of certainty that corresponds to being more likely true than not. The level of certainty would hold the determination to be at a level greater than 50 percent. This level of certainty is required for a fire cause to be classified as natural, accidental, or incendiary.
   - Possible: A level of certainty regarded as feasible but not probable. In the event that two hypotheses are formed of equal level of certainty, they each must be considered possible. When the level of certainty is determined to be possible, the cause of the fire should be considered undetermined. (page 248)
5. Oxygen in the air serves as the oxidizing agent, or oxidant. Commonly encountered supplemental oxidants include medical oxygen (e.g., compressed gas cylinders or oxygen generators) or certain chemicals (e.g., pool sanitizers). (page 246)

Fire Alarms

1. In any investigation of a fire cause, the investigator should consider all reasonable ignition sources and use the scientific method to test the hypothesis of origin and cause. These assessments of potential ignition sources are based on information and evidence gathered during the investigative process. In some investigations the physical remains of an ignition source may no longer be present. In fire scenes where the investigator has a clearly defined point of origin and there is no identifiable ignition source, the lack of evidence may afford a hypothesis for consideration. Cases may be as simple as a burned toilet paper roll in a school toilet stall being the only object burned. This case would allow a hypothesis that an open flame such as a match or lighter was applied to the available material. Determining the origin correctly makes the fire cause determination more reliable.

   The investigator would still need to review all information gathered to ensure that it could not support any other conclusion. A hypothesis based on the absence of physical evidence is one that should be approached with caution. In investigations where the ignition source is believed to be accidental in nature, that same precaution should be used. An example might be when an appliance is the suspected point of origin. The investigator should investigate the event and/or component(s) responsible for the fire. The fire could be the result of a failed component or combustibles too close to the operational heat. Because of the nature of some heat-producing appliances, the physical appearance may not yield information of cause. In such an instance witness statements and the lack of other ignition potentials can assist the investigator. (page 244)

Additional Activity

1. Page 248 lists six factors that should be considered in developing an ignition sequence:
   - How and when the initial fuel came to be present at the origin
   - How and when the oxidant came to be present (if unusual oxidant is involved)
   - How and when the competent ignition source came to be present
   - How and when the competent ignition source transferred heat to the initial fuel
   - What acts or omissions, and in what sequence, brought together the fuel, oxidizer, and ignition source
   - How the initial fuel ignited subsequent fuels to result in fire spread

Chapter 17: Analyzing the Incident for Cause and Responsibility

Matching

1. F (page 255)
2. C (page 255)
3. A (page 255)
4. E (page 255)
5. B (page 260)
6. D (page 260)
7. H (page 257)
8. G (page 257)

Multiple Choice

1. C (page 254)
2. A (page 254)
3. B (page 254)
4. D (page 254)
5. D (page 254)
6. C (page 255)
7. B (page 255)
8. A (page 255)
9. D (page 255)
10. C (page 255)

Fill-in

1. intent (page 255)
2. presumption (page 255)
3. undetermined (page 255)
4. physical (page 255)
5. motive (page 255)
6. suspicious (page 257)
7. causal factors (page 257)
8. failure analysis (page 257)

Vocabulary

1. **Incendiary fires:** Fires intentionally ignited under circumstances in which the person knows that the fire should not be ignited. (page 255)
2. **Natural fires:** Fires caused without direct human intervention or action, such as fires resulting from lightning, earthquake, and wind. (page 255)
3. **Responsibility:** The accountability of a person or other entity for the event or sequence of events that caused the fire or explosion, spread of the fire, bodily injuries, loss of life, or property damage. (page 254)
4. **Undetermined fire:** When the cause cannot be proven to an acceptable level of certainty. (page 255)

Short Answer

1. Arson is a term that denotes a crime and is therefore determined by judicial process. A fire classified as incendiary can have sufficient probable cause to be called arson by the law enforcement community. Depending on local laws, not all incendiary fires are arsons. (pages 256–257)
2. The four general categories are accidental, natural, incendiary, and undetermined. (page 255)
3. Responsibility for a fire or explosion incident is the accountability of a person or other entity for the event or sequence of events that caused the fire or explosion, spread of the fire, bodily injuries, loss of life, or property damage. Cause of the fire or explosion identifies the elements of a cause: the heat source, the first fuel ignited, the oxidizer, and the conditions or circumstances that allowed these components to come together and result in a fire or explosion. (page 254)
4. Factors contributing to the development and spread of fire and smoke that should be considered by the investigator in all cases include the following:
   - Code violations (such as propping open fire doors)
   - Compartmentation (construction detail or passive fire protection assemblies)
   - Detection and alarm systems (timeliness of activation or notification)
   - Fire suppression (both fixed systems and agency operations)
   - Fuel load and geometry
   - Housekeeping (poor housekeeping can result in more rapid spread and slow suppression activity)
   - Human behavior (acts or omissions)
   - Increase in hazard or change in occupancy (rendering original fire protection system design inadequate)
   - Structural or system failure (lack of fire stops, utility failure)
   - Ventilation effects (door and window openings, HVAC operation, and firefighting operations) (page 257)

5. Some types of analytical tools for failure analysis are as follows:
   - Timelines
   - Systems analysis
   - Failure mode and effects analysis (FMEA)
   - Fault trees
   - Mathematical modeling
   - Heat transfer analysis
   - Flammable gas concentrations
   - Hydraulic analysis
   - Thermodynamic chemical equilibrium analysis
   - Structural analysis
   - Egress analysis
   - Fire dynamics analysis (page 257)

Fire Alarms

1. Failure analysis considers what factors contributing to any injuries, loss of life, and property damage may have resulted from the root incident. The information gathered through a failure analysis of an incident can prove invaluable in helping to ensure that these losses are not repeated in the future. When looking at an incident from a failure analysis point of view, it is important to understand that it is often a chain of events that contribute to the cause.

   Some types of analytical tools for failure analysis are as follows:
   - Timelines
   - Systems analysis
   - Failure mode and effects analysis (FMEA)
   - Fault trees
   - Mathematical modeling
   - Heat transfer analysis
   - Flammable gas concentrations
   - Hydraulic analysis
   - Thermodynamic chemical equilibrium analysis
   - Structural analysis
   - Egress analysis
   - Fire dynamics analysis (page 257)

2. Failure analysis can be applied during an investigation as follows:
   - Did the sprinkler system operate? If so, why did it not control the fire? Was sufficient water provided for the hazard being protected?
   - Were there sufficient exits for people to escape from the fire? How did people react when they tried to escape?
   - Did the fire alarm sound? If not, why not? Were the smoke detectors properly placed to react to the fire development? (page 257)

Chapter 18: Failure Analysis and Analytical Tools

Matching

1. I (page 265)
2. E (page 264)
3. A (page 265)
4. B (page 265)
5. H (page 266)
6. J (page 264)
7. D (page 266)
8. G (page 269)
9. C (page 265)
10. F (page 268)

Answer Key

Multiple Choice

1. B (page 264)
2. C (page 265)
3. C (page 265)
4. A (page 265)
5. B (page 265)
6. D (page 265)
7. C (page 265)
8. B (page 266)
9. D (page 269)
10. A (page 269)

Fill-in

1. thermodynamics analysis (page 269)
2. structural analysis (page 270)
3. critical question (page 270)
4. origin and cause hypothesis (page 270)
5. computer models (page 270)
6. Specialized fire dynamics routines (page 270)
7. graphics (page 271)
8. heat release rate (page 271)

Vocabulary

1. **Benchmark event:** Event that is particularly valuable as a foundation for the timeline or may have significant relation to the cause, spread, detection, or extinguishment of a fire. (page 265)
2. **Estimated time:** An approximation based on information or calculations that may or may not be relative to other events or activities. (page 265)
3. **Failure mode and effects analysis (FMEA):** A technique used to identify basic sources of failure within a system and to follow the consequences of these failures in a systematic fashion. (page 268)
4. **Fault tree:** A logic diagram used to analyze a fire or explosion. Also known as a decision tree. (page 266)
5. **Hard time:** Identifies a specific point in time that is directly or indirectly linked to a reliable clock or timing device of known accuracy. (page 264)
6. **Heat transfer model:** Allows the investigator to determine how heat was transferred from a source to a target by one or more of the common heat transfer modes: conduction, convection, or radiation. (page 269)

Short Answer

1. The investigator should obtain data such as the location of the exits, egress routes, travel distance, and egress route widths to help in this analysis. (page 270)
2. These models may be useful in the prediction of several fire-related factors and/or events, such as time to flashover, gas temperatures or concentrations, flow rates of fire-related gases, temperatures of interior surfaces, time of activation of fire detection and suppression devices, and effects of events such as opening doors and windows. (page 270)
3. A number of variables or uncertainties that can influence fire modeling results include fire load characteristics, ventilation openings (size and open or closed), HVAC flow rates, and heat release rates. (page 270)
4. These equations and hand calculations can be used to evaluate specific issues, such as time to flashover, heat flux, heat release rate, time to ignition, flame height, detector activation, gas concentration, and flow rates of smoke, gas, and unburned fuels. (page 270)
5. Zone models usually divide a compartment into two zones, including a hot upper zone and a cooler lower zone. CFD models are more complex than zone models and divide the compartment into many small cells. (page 270)

Fire Alarms

1. A timeline is a graphic or narrative representation of events related to the fire incident, arranged in chronological order. Understanding the timeline of any incident is key to creating the failure analysis because it assists the investigator in determining the sequence of events that occurred. A timeline is a comprehensive listing of specific events, no matter how inconsequential, that can be verified to a stated degree.

    The actual timeline can be either a graphic or narrative representation of all or some activities related to a fire incident that have been arranged and formatted in chronological order. Timelines may include events that occurred before, during, and after the fire. Estimates of fire size or conditions are often valuable in developing timelines. With the use of fire dynamics, fire conditions can be related to specific events. Information gathered from detection and suppression systems may be useful in determining the spread of the fire as well as establishing a time for the events. Remember that the value of the timeline is directly related to the accuracy of the information it includes, and the reliability of a timeline is a function of the level of confidence that can be placed in specific elements that it includes. Not all information can be related with a specific time and may have to be listed as a time interval during which the event occurred.

A variety of components are used to develop timelines. These include incidents described as either hard times or soft times.

When developing a timeline, incidents that can be related to a known exact time are generally referred to as hard times. This means that the time of occurrence is specifically known. For example, a fire department's incident history records can provide the specific times when units were dispatched, arrived on scene, and so forth. Such data serve as benchmarks in developing the timeline.

Other times can be considered soft times. Soft time is either estimated or relative and is generally provided by witnesses. For example, a witness might not be able to state precisely the time a certain event such as a flashover occurred or the time when an occupant left the building but might be able to relate such observations to another event, such as the arrival time of the fire department or the moment when a bus or train went by. Through interviews and further gathering of data, an investigator may be able to narrow down the range of occurrence for such soft times to relative time periods. Relative time can be subjective in nature and varies with the witness providing the information. Witnesses should refer to their actions and observations in relation to each other, in relation to other events, and they should be as specific as possible. Estimated time is an approximation based on information or calculations that may or may not be relative to other events or activities.

Some events are particularly valuable as a foundation for the timeline or may have significant relation to the cause, spread, detection, or extinguishment of a fire. These areas are referred to as benchmark events. It is important when developing a timeline that includes different hard time sources to note any discrepancies between the clocks and synchronize the times. Such data should be double checked to ensure that errors were not made in synchronizing all such sources of data. (pages 264–265)

2. To conduct valid modeling and testing, it is important that the investigator gather data that are as accurate and complete as possible. As in all other endeavors, the "garbage in, garbage out" concept applies here. It should be anticipated that during court proceedings, test results will be subjected to a Daubert review, and the results that are presented will be only as valid as the accuracy of the information from which they were derived and the care that was taken to develop them.

Important information to be obtained includes structural information, materials and contents, and ventilation information. (page 271)

Chapter 19: Explosions

Matching

1. C (page 278)
2. E (page 279)
3. I (page 293)
4. A (page 279)
5. J (page 293)
6. B (page 280)
7. D (page 279)
8. F (page 280)
9. G (page 278)
10. H (page 279)

Multiple Choice

1. B (page 278)
2. C (page 278)
3. B (page 278)
4. A (page 280)
5. B (page 282)
6. A (page 283)
7. D (page 283)
8. C (page 283)
9. D (page 283)
10. A (page 284)
11. B (page 284)
12. C (page 285)
13. A (page 286)
14. D (page 287)
15. B (page 288)

Fill-in

1. chemical explosions (page 279)
2. deflagration (page 279)
3. time (page 279)
4. low-order damage (page 280)
5. seismic effect (page 283)
6. amplifying (page 283)
7. seat of the explosion (page 283)
8. confining space (page 285)
9. migrate underground (page 286)
10. dust (page 286)

Vocabulary

1. **Burning velocity:** The rate of flame propagation relative to the velocity of the unburned gas ahead of it. (page 284)
2. **Deflagration:** A reaction that propagates at a subsonic velocity, several feet per second, and can be successfully vented. (page 279)
3. **Explosion dynamics analysis:** The process of using force vectors to trace backward from the least to the most areas following the general path of the explosion force vectors. (page 290)
4. **Firebrands:** Hot or burning fragments propelled from an explosion. (page 283)
5. **High-order damage:** A rapid pressure rise or high force explosion characterized by a shattering effect on the confining structure or container and long missile distances. (page 280)
6. **Mechanical explosion:** That of high-pressure gas producing a physical reaction such as the rupture of a container. The fundamental nature of the fuel is not changed. (page 278)
7. **Seismic effect:** The transmission of tremors through the ground as a result of the blast wave expansion, causing structures to be knocked down. (page 283)
8. **Transitional velocity:** The sum of the velocity of the flame front caused by the volume expansion of the combustion products due to the increase in temperature and an increase in the number of moles (a mole is the quantity of anything that has the same number of particles found in 12,000 grams of carbon-12) and any flow velocity caused by motion of the gas mixture before ignition. (page 284)

Short Answer

1. Explosions have four major effects:
   - Blast overpressure and wave effect
   - Projected fragments effect (shrapnel)
   - Thermal effect
   - Seismic effect (ground shock) (page 280)
2. The shape of the blast front from an idealized explosion is spherical, expanding evenly in all directions from the epicenter. In the real world, confinement, obstruction, ignition position, cloud shape, or concentration distribution at the source of the blast pressure wave changes and modifies the direction, shape, and force of the front. (page 282)
3. The factors that control the effects of an explosion are as follows:
   - Type and configuration of fuel
   - Nature, size, volume, and shape of containment vessel
   - Level of congestion and obstacles within the vessel
   - Location and magnitude of ignition source
   - Venting of the containment vessel
   - Relative maximum pressure
   - Rate of pressure rise (page 283)
4. The explosion damage to structures is related to several factors:
   - Fuel-to-air ratio
   - Vapor density of the fuel
   - Turbulence effects
   - Volume of the confining space
   - Location and magnitude of the ignition source
   - Venting
   - Strength of the structure (page 284)
5. The investigator should make an initial assessment of the type of incident. If the investigator determines that the explosion was fueled by explosives or an explosive device, the investigator should discontinue the scene investigation, secure the area, and contact the appropriate entities. (page 288)

Fire Alarms

1. The investigator should make an initial assessment of the type of incident. If the investigator determines the explosion was fueled by explosives or an explosive device, the investigator should discontinue the scene investigation, secure the area, and contact the appropriate entities. The following tasks are sequenced to assist the investigator in the initial scene assessment:
   - Identify whether the incident was an explosion or fire
   - Determine high- or low-order damage
   - Identify seated or nonseated explosion
   - Identify the type of explosion
   - Identify the potential general fuel type
   - Establish the origin
   - Establish the fuel source and explosion type
   - Establish the ignition source (page 288)

2. Identification of the fuel source begins by determining the types of fuel that were available at the site through locating and then inspecting utility services (especially fuel gas), production by-products such as dusts or particles, and any ignitable liquids present for any reason. Damage at the scene is compared with typical damage patterns of lighter-than-air gases, heavier-than-air gases, liquid vapors, dusts, explosives, backdrafts, and BLEVEs. (page 289)

Additional Activity

1. Page 287 has three very good paragraphs explaining differences between low and high explosives as well as identifying some common explosives.

   Low explosives are characterized by deflagration (subsonic blast pressure wave), a slow rate of reaction with the development of low pressure. Examples of low explosives are smokeless gunpowder, flash powders, solid rocket fuels, and black powder. Low explosives are designed to work by pushing or heaving effects.

   High explosives are designed to produce shattering effects because of their high rate of pressure rise and extremely high detonation pressure. Thus, a characteristic of high explosives is the detonation propagation mechanism. The high, localized pressures are responsible for creating localized damage near the center of the explosion.

   A deflagration (fuel–air) explosion usually results in structural damage that is uniform and omnidirectional with relatively widespread evidence of burning, scorching, and blistering. The rate of combustion of a solid explosive is extremely fast compared with the speed of sound. Thus, pressure does not equalize throughout the explosion volume, and high pressures are generated near the explosive. The location of the explosion should be evidenced by the crushing, splintering, and shattering that are produced by the higher pressure; however, major distance from the source of the explosion may leave little evidence of intense burning or scorching except where shrapnel may have landed on combustible materials. Common examples of high explosives are dynamite, water gel, TNT, ANFO, RDX, and PETN. The extremely high detonation pressure may reach 1,000,000 psi (6,900,000 kPa).

Chapter 20: Incendiary Fires

Matching

1. D (page 302)
2. F (page 305)
3. J (page 305)
4. A (page 299)
5. I (page 305)
6. H (page 305)
7. B (page 305)
8. C (page 305)
9. G (page 305)
10. E (page 298)

Multiple Choice

1. C (page 305)
2. D (page 305)
3. C (page 302)
4. A (page 305)
5. B (page 305)
6. B (page 305)
7. A (page 301)
8. B (page 303)
9. D (page 300)
10. C (page 305)

Fill-in

1. incendiary fire (page 298)
2. Multiple fires (page 298)
3. trailer burn (page 299)
4. heat sources (page 299)
5. fire ignition (page 300)
6. location (page 302)
7. accidental (page 302)
8. place obstructions (page 302)
9. tampering (page 302)
10. Extremism (page 306)

Vocabulary

1. **High-temperature accelerant (HTA):** Mixtures of fuels with Class 3 or Class 4 oxidizers and thermite mixtures. (page 299)
2. **Incendiary device:** A wide range of mechanisms used to initiate an incendiary fire. (page 300)
3. **Personation:** The unusual behavior by an offender, beyond that necessary to commit the crime. (page 305)
4. **Spree arson:** The setting of three or more fires at separate locations with no emotional cooling-off period between fires. (page 305)
5. **Staging:** Purposeful alteration of the crime scene before the arrival of police. (page 305)
6. **Trailer:** The method to spread fire to other areas by deliberately linking them together with combustible fuels or ignitable liquids. (page 299)
7. **Vandalism:** Mischievous or malicious fire setting that results in damage to property. (page 305)
8. **Victimology:** A thorough understanding of the offender activity with the victim (or targeted property). (page 305)

Short Answer

1. "Natural" means of fire spread that could cause multiple fires are
   - Conduction, convection, or radiation
   - Flying brands
   - Direct flame impingement
   - Falling flaming materials (drop down)
   - Fire spread through shafts
   - Fire spread within wall or floor cavities
   - Overloaded electrical wiring
   - Utility system failures
   - Fuel gas or dust explosions
   - Lightning (page 298)
2. Materials that can be used as trailers are
   - Ignitable liquids
   - Clothing
   - Paper
   - Straw (page 299)
3. Indicators of high-temperature accelerants (HTA) are
   - Rapid rate of growth
   - Brilliant flares
   - Melted steel or concrete (page 299)
4. Some examples of an incendiary device include (but are not limited to) the following:
   - Combination of cigarette and matchbook
   - Candles

- Wiring systems
- Electric heating appliances
- Fire bombs/Molotov cocktails
- Paraffin wax–sawdust incendiary device (fireplace starters) (page 300)

5. Six motive classifications determined to be the most effective in identifying offender characteristics for fire-setting behavior are
   - Vandalism
   - Excitement
   - Revenge
   - Crime concealment
   - Profit
   - Extremism (page 305)

Fire Alarms

1. Some incendiary devices are constructed as delay devices to allow the fire-setter to leave the area safely. If, during the investigation, the investigator finds a device that has not been activated, he or she should not move it. Adequate precautions and safeguards should be taken, including the notification of trained explosive ordnance disposal personnel if an active or live device is found. (page 300)
2. There may be occasions when the fire-setter tries to mask the true cause of the fire by attempting to use an appliance as the "obvious" cause of the fire. For example, a fire-setter might pour an ignitable liquid into a coffee maker, causing a fire to occur. The investigator should not assume the appliance malfunctioned, causing the fire. Further testing or evaluation may be warranted to determine the true cause of the fire. The investigator should be aware of spoliation issues when conducting this type of investigation. Caution should be taken to protect these mechanisms until all interested parties have been put on notice and are present before any destructive testing is done. An alternative to destructive testing is the use of an X-ray machine to document the contact positions of the controllers without taking the panel apart.

   Because arsonists may set multiple fires, the investigator should inspect the building to determine whether other fires occurred that are noncommunicating. The arsonist might have used similar devices in these other fires that can provide valuable clues. Examine these areas for debris that may contain delay or other incendiary devices or methods that would contribute to the ignition sequence. In cases of multiple fires, often at least one incendiary device has failed to operate, leaving the investigator with valuable evidence. Look for trailers from one burn area to the other(s). (pages 300–301)

Chapter 21: Fire and Explosion Deaths and Injuries

Matching

1. C (page 314)
2. H (page 318)
3. F (page 312)
4. A (page 315)
5. G (page 316)
6. B (page 313)
7. D (page 314)
8. E (page 319)

Multiple Choice

1. B (page 312)
2. A (page 312)
3. C (page 313)
4. D (page 313)
5. A (page 313)
6. C (page 314)
7. A (page 314)
8. B (page 314)
9. D (page 316)
10. C (page 316)

Fill-in

1. stability (page 316)
2. 75; 80 percent (page 316)
3. Hyperthermia (page 318)
4. edema (page 318)
5. Hypoxia (page 318)
6. fatal investigations (page 319)
7. conducted heat (page 320)
8. blast pressure, shrapnel, (page 321) thermal, seismic

Vocabulary

1. **Carboxyhemoglobin (COHb):** The carbon monoxide saturation in the blood. (page 316)
2. **Hypoxia:** Condition caused by a victim breathing in a reduced oxygen environment. (page 318)
3. **Lividity:** Occurs after death and is the pooling of the blood in the lower elevations of the body caused by the effects of gravity. (page 315)
4. **Pugilistic attitude:** A crouching stance with flexed arms, legs, and fingers. (page 315)

Short Answer

1. Diagrams should detail and record the physical dimensions of the scene, the contents of the scene, and the measurements of the body location. (page 313)
2. In cases of death the investigation expands the team to possibly include a homicide detective, the medical examiner, the coroner, forensic laboratory personnel, and a forensic pathologist. If the body is badly burned, the expertise of a forensic anthropologist and a forensic odontologist (dentist) is likely to be required. (page 313)
3. Burn injuries need to be documented and assessed to the following degree of burns:
   - First degree: Reddening of the skin; also called *superficial* burn
   - Second degree: Blistering of the skin; also called *partial-thickness* burn
   - Third degree: Full-thickness damage to the skin, also called *full-thickness* burn
   - Fourth degree: Damage to the underlying tissue and charring of the tissue (page 319)
4. The location and distribution of explosion injuries may indicate location and activity of the victim at the time of the explosion and may help to establish the location, orientation, energy, and function of the exploding mechanism or device. (page 321)
5. Various tests and documentation during the postmortem examination provide the investigator with valuable information to aid in identifying the victim as well as establishing the cause and manner of the death. These tests may include the following:
   - Blood: For level of COHb, HCN concentration, drugs, alcohol, or poisons
   - Internal tissue: For level of volatile hydrocarbons, drugs, or poisons
   - Stomach: For contents, including the presence or absence of soot
   - Airways: For effects of the fire
   - Internal body temperature: To assist in establishing a time and mechanism of death
   - X-rays: To assist in identification or location of foreign objects in the body
   - Clothing or personal effects: To assist in identification or check for the presence of ignitable liquids
   - Recovery of foreign objects in or on the body: Possibly to include bullets, knife parts, explosive device components, and other items recovered
   - Sexual assault evidence: To provide possible motivation for setting a fire
   - Documentation by photography and sketching: To record injuries and burns to the body and evidence recovered from the body (page 319)

Labeling

1. The rule of nines. (page 320)

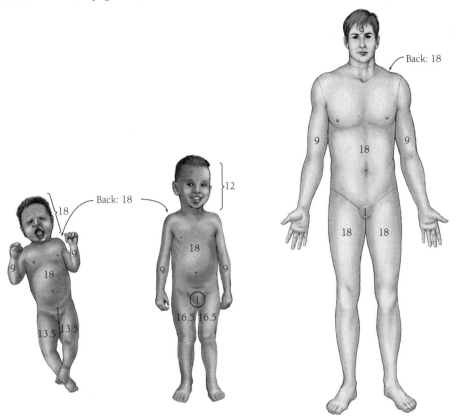

Fire Alarms

1. All fire or explosion investigations require a team effort. In cases of death the investigation often expands the team to include a homicide detective, the medical examiner, the coroner, forensic laboratory personnel, and a forensic pathologist. If the body is badly burned, the expertise of a forensic anthropologist and a forensic odontologist (dentist) is likely to be required. (page 313)

2. Documentation of the scene should include sketches and diagrams. Diagrams should detail and record the physical dimensions of the scene, the contents of the scene, and the measurements of the body location. The sketch should include the outline of the body for reference purposes. Often, these sketches are useful in court when photos of a victim are not admissible out of consideration for jury members. The investigative process may include the creation of a victim sketch. This sketch is used to detail burn and other injuries to the victim.

   To aid in a thorough examination, investigators can use a grid system to divide the scene into sections. Each grid needs to be examined and documented, and evidence needs to be identified. If the scene does not lend itself to a grid search, the investigators might wish to consider a spiral or grid search pattern. All methods used should have some overlap for complete coverage. (page 313)

Chapter 22: Appliances

Matching

1. D (page 331)
2. F (page 333)
3. J (page 333)
4. I (page 330)
5. H (page 337)
6. E (page 340)
7. A (page 333)
8. G (page 340)
9. B (page 335)
10. C (page 330)

Multiple Choice

1. C (page 340)
2. A (page 326)
3. D (page 328)
4. A (page 328)
5. B (page 330)
6. C (page 330)
7. D (page 330)
8. B (page 330)
9. C (page 331)
10. A (page 331)

Fill-in

1. were connected to (page 332)
2. energy density (page 332)
3. switch (page 333)
4. motion (page 334)
5. surrounding combustibles (page 336)
6. fluorescent (page 336)
7. short circuits (page 336)
8. arc (page 337)

Vocabulary

1. **Arc tube:** A cylinder of fused silica/quartz with electrodes at either end that is used in most mercury and metal halide lamps. (page 337)
2. **Exemplar:** An exact duplicate of an appliance. (page 330)
3. **Radiation heater:** A heater that transfers heat by radiation only. (page 340)
4. **Step-down transformer:** Device that reduces the 120 V provided at the receptacle to the required voltage. (pages 331–332)
5. **Transformer:** Device that reduces AC voltage, usually 120 or 240 V AC, to a lower voltage and can be used to isolate the appliance from its power source. (page 335)
6. **Triac:** A semiconductor switch that conducts current when the gate is turned on and blocks current when the gate is turned off. (page 335)

Short Answer

1.

Sensor Name	Property Sensed	Sensor Type, Output	Typically Controls
Thermocouple	Temperature	Dissimilar metal contact produces millivolt signal	Heat-producing appliances
Flame sensor	Presence of flame	Infrared detector, microvolt signal	Gas valve, blower motor
Current sensor	Current	Current passes through series resistance causing voltage drop, or magnetic measurement causes change in resistance	Current-controlled heaters or cooking appliances, variable speed motors

2. Major causes of dishwasher fires are
   - Moisture contacting the conductors, especially at the top of the door where the controls are.
   - Repeated opening and closing of the door over time can stress the wiring harness. The insulation breaks down and a short circuit can result.
   - A combustible material contacting the electrical elements could also cause a fire.
   - Some control modules in dishwashers contain relays that have failed and self-ignited the plastic housing, causing a fire. (page 340)
3. The three conditions are
   - Combustible material getting caught
   - The heating element detaching because of rough handling and later igniting the plastic casing
   - Restriction of the inlet air, but not the overtemperature sensor, causing the heating element to overheat and ignite the plastic casing (page 340)
4. A clogged lint filter can blow lint back into the interior of the dryer. The lint can then settle across the base, eventually reaching the element and igniting. Frictional heating from a piece of clothing caught between the moving parts may cause a fire, and the heating of contents such as vegetable oil–soaked rags and plastic bags may cause them to spontaneously combust. (page 342)
5. In a fire scene the most convincing indicators that an appliance caused the fire are as follows:
   - Fire patterns that point to the area of origin being near the appliance.
   - Severe fire damage to the appliance.
   - Arcing or melting found on conductors either inside or near the appliance. (page 343)

Labeling

1. Refrigerator components. (page 339)

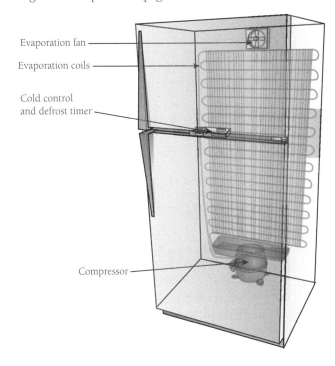

2. Clothes dryer components. (page 341)

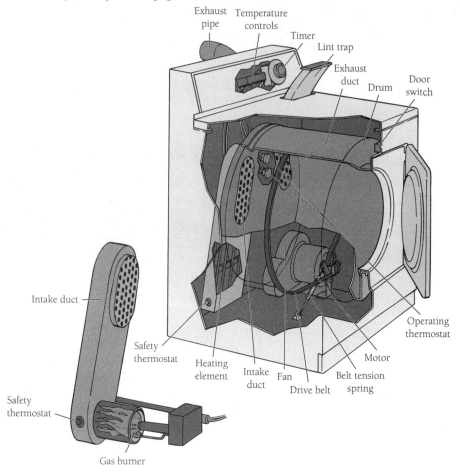

Fire Alarms

1. In a fire scene the most convincing indicators that an appliance caused the fire are as follows:
   - Fire patterns that point to the area of origin being near the appliance
   - Severe fire damage to the appliance

   In addition, if the appliance operates on electrical power:
   - Arcing or melting found on conductors either inside or near the appliance

   Water-using appliances such as dishwashers, washing machines, water heaters, and outdoor appliances:
   - Consider fire caused by water/moisture or the lack of it. (pages 326 and 330)

2. Fluorescent lighting systems are commonly used in office settings. They use one or more glass tubes filled with a starting gas and low-pressure mercury. An electrical discharge is sent down the length of the glass tube, exciting the mercury gas. Two methods of starting fluorescent lights are used: the ballast discharge system, which is more common, and the preheated filament system. Ballasts are typically transformers (sometimes called magnetic), but electronic ballasts are becoming more popular, especially in compact fluorescent lamps (CFLs), and work on a switched transistor principle. The ballast is used for creating the high start-up discharge voltage required to establish a starting arc across the lamp electrodes. This process is enabled by the starting gas, which results in the mercury vapor creating ultraviolet light that is converted to visible light by the coating on the inside of the tube, known as the phosphor or fluorescent powder. Once the current has started flowing and the lamp is operating, the ballast acts to limit the current through the tube.

   There are two main types of fluorescent light ballasts: magnetic and electronic. The ballast in fluorescent lights can overheat because of internal short circuiting and ignite combustible ceiling materials. Magnetic ballasts incorporate either a reactor or a transformer. Interior fluorescent light fixtures manufactured after 1968 are required to have thermal protection in the ballast.

A "P" on most metal housings indicates that thermal protection exists. All fixtures, indoor and outdoor, manufactured after 1990 are required to have thermal protection in the ballast; however, thermal protection does not ensure that a failure cannot occur. Both the electronic and magnetic fluorescent light ballasts contain pitch or potting compound within the ballast. This provides better heat transfer, reduces noise, and holds the internal parts in place. This pitch can ooze out because of either internal heating or fire exposure. This pitch will normally not ignite other materials unless it is already burning. The electronic ballasts in CFLs use semiconductor switches (transistor/Mosfet/triac) and relatively small passive circuits (capacitor/inductor) switching at high frequency to create the electrical arc required. Some electronic ballasts use self-resetting thermal protectors, whereas others use fuses for thermal protection. The ballast-control insulation contacts (ICs) usually contain overcurrent protection, which must work correctly for the switching action to be present and for these ballasts to work correctly. For an unsafe condition to occur, the IC must malfunction (many ICs have protection against this as well), and the semiconductor switches must short circuit to generate significant heat.

Common failures of magnetic ballasts that initiate fires include arc penetrations into combustible ceiling materials or nearby combustibles and extreme coil overheating that conduct heat into nearby combustibles. If the investigator suspects a ballast failure as the cause of a fire, the ballast along with the fixture and the wiring should be preserved for examination by qualified experts. The fixture itself along with conductors feeding power to it should be examined for failures, such as arcing nearby or inside the fixture and failure of the lamp holders. (page 336)

Chapter 23: Motor Vehicle Fires

Matching

1. H (page 357)
2. E (page 363)
3. A (page 349)
4. C (page 357)
5. B (page 354)
6. G (page 349)
7. D (page 355)
8. F (page 361)
9. J (page 349)
10. I (page 355)

Multiple Choice

1. A (page 349)
2. C (page 354)
3. D (page 355)
4. B (page 354)
5. D (page 356)
6. A (page 356)
7. C (page 357)
8. B (page 359)
9. D (page 360)
10. A (page 364)

Fill-in

1. propane; natural gas (page 349)
2. ethanol (page 349)
3. ignitable liquid (page 351)
4. solid magnesium (page 351)
5. structure fires (page 351)
6. not running (page 351)
7. electrical arcing (page 353)
8. exhaust system (page 353)
9. natural gas; LP gas (page 356)
10. better accomplished (page 360)

Vocabulary

1. **Catalytic converter:** A device in the exhaust system that exposes exhaust gases to a catalyst metal to promote oxidation of hydrocarbon materials in the exhaust gas. (page 354)
2. **Electronic (or engine) control module (ECM):** An electronic device that controls engine operation parameters, including fuel delivery, throttle control, and safety systems operations. (page 357)
3. **Event data recorder (EDR):** A device to record data before and after a crash event. (page 357)

Short Answer

1. The following factors all affect ignition of liquids by a hot surface:
   - Ventilation
   - Environmental factors
   - Autoignition point

- Liquid flash point
- Liquid boiling point
- Liquid vaporization rate
- Atomization of the liquid (effective fuel surface area)
- Latent heat of vaporization of the liquid
- Length of exposure of the liquid to the heated surface and configuration of the hot surface (page 354)

2. Information that may be helpful to obtain includes the following:
   - Last use of the vehicle and by whom
   - Mileage at the time of the fire
   - Operation or problems
   - Service or maintenance history
   - Fuel level and type, when last fueled and where
   - Equipment
   - Personal effects in each area of the vehicle
   - Photos or videos before, during, or after the fire (page 359)

3. RVs may be equipped with equipment and systems unlike those found in other vehicles, such as the following:
   - Shore power connections: These consist of a cord, sometimes with a collection of adapters, for connection to a nearby receptacle, as may be available at an RV park or at the owner's home.
   - Auxiliary power generator systems: These are either fixed or portable and are powered by an internal combustion engine fueled by gasoline, diesel, or propane. Incorporated may be an automatic generator starting system (AGS) that automatically starts and stops the generator under preset conditions of load, demand, battery condition, or time.

   Electrical converters and inverters: These allow power to be interchanged between 12-V DC and 120-V AC systems. The converter transforms 120-V AC power to 12-V DC, whereas the inverter performs the opposite function. Often, these functions are combined in one electronically controlled unit, which also may be connected to the coach and chassis batteries as an automatic charger.

   Battery systems: Most often, the self-propelled RV uses two separate battery systems. The chassis has one or more 12-V DC batteries for starting the engine and operating other vehicle systems such as exterior lights and heating systems. The coach typically uses a separate set of two or more 12-V DC batteries to provide 12-V power to living spaces. The two sets of batteries are often connected through an electronic isolator system to allow the batteries to be evenly charged by the engine generator, the auxiliary generator, or the converter system.

   - Holding tanks: The RV may be equipped with multiple holding tanks for toilet waste (black water), sink and shower waste (grey water), and fresh or potable water.
   - Propane systems: Compressed propane stored in one or more tanks on the RV supplies fuel to power the range or cooktop, refrigerator, water heater, and furnace if so equipped. The pressure is modified by a two-stage regulator and fuel is delivered through pipes or tubing to the individual appliances.
   - Water heaters: Powered by propane, 12-V AC, or a combination of both. Typically, 12-V DC power is used for control circuits, using an internal thermostat to regulate water temperature.
   - Furnaces: Typically forced-air, using propane as the fuel, with controls and blowers operating on 12-V DC.
   - Refrigerators: RV refrigerators do not use a compressor, like home refrigerators; rather, their design involves a closed system of tubes containing water, anhydrous ammonia, sodium chromate, and hydrogen gas under pressure. The ammonia/water mixture is heated by a small propane burner, a 120-V AC heating element, or a 12-V DC heating element. Automatic controls may switch between fuel sources, or the operator may manually select one mode of operation.

   Historically, fires involving RV refrigerators have involved loss of coolant containment (ammonia and/or hydrogen gas), electrical problems with control circuits, or venting restrictions. (page 362)

4. Diesel engines are still standard in most buses, but some districts are now using CNG, LNG, diesel electric hybrid, or pure electric drive systems. (page 363)

5. The drive system may use up to 600 volts, supplied by a battery pack, whereas peripheral systems such as lighting, sound, and other convenience systems use a traditional automotive-style 12-V DC system supplied from a standard 12-V battery. (page 364)

Fire Alarms

1. In most cases the sources of ignition energy in vehicles are the same as those associated with structure fires. Open flames, mechanical and electrical failures, and discarded smoking materials can often lead to a vehicle fire. Vehicles do have some unique ignition sources, which should be considered and evaluated by the investigator, including heated exhaust components, various types of bearings, and braking systems.

   Exhaust system components can provide sufficient temperature to ignite diesel spray and to vaporize gasoline. Slippage of transmission or torque converter components may result in sufficient heat being generated to vaporize fluid, causing a discharge of fluid that can ignite on contact with heated exhaust system components.

   Engine oil, power steering fluid, and brake fluid can also ignite when in contact with heated exhaust system components. These liquids may ignite during operation or soon after the engine is shut off, when the temperature of some exhaust system components is still sufficiently elevated.

   Catalytic converters normally operate with an internal temperature approaching 1300°F (700°C) and a surface temperature in the vicinity of 600°F (315°C) but may reach higher temperatures under heavy loads if air circulation is restricted or if unburned fuel enters the converter because of overfueling or engine misfiring. Under such extreme circumstances, the ceramic matrix material inside the converter may melt and may be ejected from the tailpipe in an incandescent spray.

   Another ignition source identified in vehicle fires is improperly discarded or misused smoking materials. (pages 351, 353, and 354)

2. It is advantageous to document the vehicle at the fire scene using the same procedures used for documenting a structural fire scene. The investigator should start with the documentation of the area surrounding the vehicle and then focus on the vehicle itself. When possible, the position and location of the vehicle and its relationship to buildings, streets, and other site features should be documented photographically and by measuring and diagramming the site.

   It is advisable to determine the actions and events that led up to the fire by conducting interviews with the person(s) who first observed the fire. The owner, passengers, and operator of the vehicle at the time of the fire or the last person to drive the vehicle; nearby residents; fire department personnel; and police officers might also provide pertinent information. Information that may be helpful to obtain includes the following:

   - Last use of the vehicle and by whom
   - Mileage at the time of the fire
   - Operation or problems
   - Service or maintenance history
   - Fuel level and type, when last fueled and where
   - Equipment
   - Personal effects in each area of the vehicle
   - Photos or videos before, during, or after the fire

   The investigator should record the fire scene by making a scene diagram showing reference points and distances. The scene should be photographed, as should surrounding buildings, highway structures, vegetation, other vehicles, tire and foot impressions, fire damage, signs of fuel discharge, and any parts or debris.

   The vehicle should be photographed at the scene in a systematic manner similar to that for a structure fire. The investigator should begin on the outside, documenting all surfaces (including top and underside, if possible), damaged and undamaged areas, tires and tire tread depth. The investigator should document the engine compartment, taking overview photographs of all sides and focusing on specific engine areas and components. The investigator should determine and document the positions of windows and doors, inspecting window lift mechanisms for their positions and side window channels for evidence of remaining glass. The investigator should document the paths of fire spread into or out of compartments and cargo spaces. Photos of the cargo space should include photographs of the spare tire and any special equipment such as stereo gear or add-on devices when necessary.

   The investigator should photograph any changes or alterations made to the vehicle during the fire suppression effort and/or the initial inspection and, when possible, should also document the vehicle removal process. (pages 359–360)

Chapter 24: Wildfire Investigations

Matching

1. D (page 382)
2. E (page 372)
3. G (page 380)
4. A (page 370)
5. B (page 382)
6. I (page 380)
7. J (page 370)
8. C (page 371)
9. F (page 373)
10. H (page 372)

Multiple Choice

1. C (page 370)
2. A (page 370)
3. B (page 370)
4. B (page 371)
5. C (page 371)
6. A (page 371)
7. D (page 371)
8. D (page 371)
9. B (page 371)
10. A (page 373)
11. C (page 373)
12. B (page 373)
13. D (page 373)
14. C (page 380)
15. B (page 380)

Fill-in

1. ground fuels; surface fuels; aerial fuels (page 370)
2. curing (page 370)
3. lower level fuels (page 371)
4. higher; slower (page 372)
5. areas of origin (page 373)
6. char pattern (page 374)
7. lane technique (page 379)
8. fulgurites (page 380)
9. incendiary fire (page 380)
10. prescribed fire (page 380)

Vocabulary

1. **Aerial fuel:** All green or dead materials located in the forest canopy. (page 370)
2. **Aspect:** The direction the slope faces (N, E, S, W). (page 372)
3. **Crown:** Twigs, needles, or leaves of a tree or bush. (page 371)
4. **Cupping:** A charred surface on the fuel that is caused by the exposure of the surface to the windward side of the fire. (page 375)
5. **Duff:** The layer of decomposing organic materials lying below the litter layer of freshly fallen twigs, needles, and leaves and immediately above the mineral soil. (page 370)
6. **Fire head:** The portion of a fire that is moving most rapidly, subject to influences of slope and other topographic features. Large fires burning in more than one drainage of fuel type can develop additional heads. (page 371)
7. **Fire heel:** Located at the opposite side of the fire from the head, this part of the fire is less intense and is easier to control. (page 371)
8. **Fire storm:** A natural phenomenon that attains such intensity it creates and sustains its own wind system. (page 373)
9. **Firing out:** The process of burning the fuel between a fire break and the approaching fire to extend the width of the fire barrier. (page 373)
10. **Fire flanks:** The parts of a fire's perimeter that are roughly parallel to the main direction of spread. (page 373)
11. **Fulgurites:** A slender, usually tubular, body of glassy rock produced by electrical current striking and then fusing dry sandy soil. (page 380)

Short Answer

1. Fuels are broken down into four major fuel groups: grass, shrub, timber litter, and logging debris. (page 371)
2. The primary factors that affect fire spread are heat transfer, lateral confinement, weather influence, fuel influence, suppression efforts, and topography. (page 371)

3.

Direction	Conditions	Effect
Facing toward the sun	• Has less vegetation • Dryer	• Greater ease of ignition and fire spread
Facing away from the sun	• Has more vegetation • More moist	• Less ease of ignition and fire spread

(page 372)

4. Visual indicators include differential damage; char patterns; discoloration; carbon staining; and the shape, location, and condition of residual, unburned fuel. (page 373)
5. The basic principles of safety apply for wildfire incidents. These principles include the following 10 standard firefighting orders:
   - Keep informed on fire weather conditions and forecasts.
   - Know what your fire is doing at all times.
   - Base all actions on current and expected behavior of the fire.
   - Identify escape routes and safety zones, and make them known.
   - Post lookouts when there is possible danger.
   - Be alert. Keep calm. Think clearly. Act decisively.
   - Maintain prompt communications with your forces, your supervisor, and adjoining forces.
   - Give clear instructions, and ensure they are understood.
   - Maintain control of your forces at all times.
   - Fight fire aggressively, having provided for safety first. (page 376)

Fire Alarms

1. Analysis of the directional pattern shown by multiple indicators in a specific area can identify the path of fire spread through the area. By using a systematic approach of backtracking the progress of the fire, the investigator can retrace the path of the fire to the point of origin. This procedure is the accepted and standard technique in wildfire investigation. Visual indicators include differential damage; char patterns; discoloration; carbon staining; and the shape, location, and condition of residual, unburned fuel.

   Wildfire V-shaped patterns are ground surface burn patterns created by the fire spread. When viewed from above, they are generally shaped like the letter V. These are not to be confused with the traditional plume-generated vertical V patterns associated with structural fire investigations.

   Wildfire V-shaped patterns have the following characteristics:
   - Horizontal, not vertical, patterns
   - Affected by wind direction and slope
   - Fire flanks (perimeter of fire parallel to direction of spread) widen up the slope in the direction of the wind
   - Origin that is normally near the base of the V

   The degree and type of damage to fuels indicate the intensity, duration, and direction of fire passage. Leaves, branches, and large woody material sustain greater damage on the side from which the fire approached. Grass that grows in clumps may not be completely consumed, resulting in protected areas on the side opposite the fire's advance.

   As a low-intensity fire burns the bottoms of grass stems, the stems fall back into the fire. If they fall into a burned area behind the fire head, they may remain unburned. Unburned stalks of grass on the ground generally point in the direction of the fire approach. The investigator should examine several grass stems to increase the reliability of this method. He or she should also view the stems outside the burned area to ensure that no other influences, such as wind or heavy rains, have had an effect on their vertical arrangement.

   Brush can be very valuable as a fire direction indicator. It is often damaged by the fire but not fully consumed, leaving the investigator a reliable indicator that has not been moved or destroyed, even during light mop-up. Brush can display several indicators such as freezing, degree of damage, depth of char, angle of char, curling of leaves, sooting, white ash deposits, cupping, and die-out patterns. Not many objects can display as many indicators as brush.

Trees are significant indicators of fire direction, particularly in areas of frontal fire damage. Fire movement is recorded at ground level around the root base and tree trunk and at flame height by the lower foliage and crown canopy. The char to the trunk surface of a tree is affected by the topography of the land surface. A fire burning uphill or with the wind creates a char pattern that slopes to a greater degree than the ground slope. Crown damage can also be used to interpret fire direction. Convection and radiant heat travel ignite lower limbs and then spread upward into the rest of the tree. This action progresses in intensity as the wind action drives the fire away from the windward foliage and branches. (pages 373–374)

2. When completing a search for a fire's origin, the following principles should be used:
   - Walk the exterior of the fire.
   - Look for burn patterns.
   - Identify the general area of origin.
   - Enter the general area of origin.
   - Identify the specific area of origin.
   - Grid the specific area of origin.
   - Identify the origin.

The search area should be walked twice in opposite directions to view burn patterns from different angles while walking the exterior of the fire. While examining burn patterns, identify clusters or groups of indicators to ensure your finding of spread direction. Once you are sure of the general area of origin, the flagged-off area can be reduced in size to allow suppression crews to work the area. The general area of origin is the area of the fire that the investigator can narrow down based on macroscale indicators, witness statements, and analysis of fire behavior. (page 378)

Chapter 25: Management of Complex Investigations

Matching

1. F (page 386)
2. D (page 386)
3. C (page 392)
4. B (page 388)
5. A (page 387)
6. E (page 388)
7. G (page 388)
8. H (page 387)

Multiple Choice

1. A (page 387)
2. C (page 388)
3. D (page 388)
4. B (page 388)
5. D (page 388)
6. A (page 388)
7. C (page 389)
8. C (page 392)
9. D (page 394)
10. B (page 394)

Fill-in

1. interested parties (page 386)
2. work plan (page 387)
3. protocol (page 387)
4. evidence altering (page 388)
5. agreements; protocols (page 391)
6. suggested methods (page 391)
7. activities and schedule (page 388)
8. scheduling (page 389)

Vocabulary

1. **Evidence custodian:** Person who is responsible for managing all aspects of evidence control. (page 392)
2. **Interested party:** Any person, entity, or organization, including their representatives, with statutory obligations or whose legal rights or interests may be affected by the investigation of a specific incident. (page 386)
3. **Protocol:** A description of the specific procedures and methods by which a task or tasks are to be accomplished. (page 388)
4. **Work plan:** An outline of the tasks to be completed as part of the investigation, including the order or timeline for completion. (page 387)

Short Answer

1. Some of the concerns to be addressed are safety, protocols, work plans, planning and timing, access, organizing the investigative team, regular meetings, resources, preliminary information, lighting, securing the scene, sanitary and comfort needs, communications, interviews, plans and drawings, search patterns, and evidence. (page 388)

2. Such cost sharing may include professional photography or videography during the investigative activities, cost in developing drawings, or site safety PPE. Sharing of cost may also include evidence storage, debris removal, personal comfort items, and specialized tests of the scene environment or evidence. (page 389)

3. This may include the owner, representatives of injuries or fatalities, and parties of potential civil and criminal prosecution as well as building component manufacturers, appliance systems providers, protection and detection system installers, equipment manufacturers, and construction material and finish manufacturers. Other parties that should not be overlooked may include service companies that serviced the equipment in the area of origin, suppression or detection equipment representatives, representatives relating to the spread of the fire, and even some public agencies such as code enforcers or inspectors. (page 387)

4. The protocol and work plan may be separate documents or combined in one document. The protocol may be a general agreement between parties of how the investigation is to proceed and the methods to be used. The work plan may be an additional document on details of each activity conducted during the investigation, the timelines for such activities, safety monitoring, and structural work. The work plan may be a more fluid document that changes as the work progresses or as issues evolve during the scene work. (page 387)

5. Regular communications may be conducted in various forms such as through regular meetings, Web sites, e-mail, or bulletin boards. (page 388)

Fire Alarms

1. Often, one of the most difficult responsibilities borne by the manager of the complex investigation is to identify the interested parties who should have access to the site. This is usually coordinated with a legal advisor who is then likely to send out the notices. The owner/occupant, the insurance carrier, and the legal representatives should be part of the identification and notification process. Generally, an interested party is one who may be involved in future litigation and may have an interest as the artifacts or evidence are removed or altered.

   Interested parties may be identified early in the investigation process. During the course of an investigation, however, other parties may be identified. These parties should be notified of the incident; they may have some responsibility, and they may participate in the investigation. Interested parties that have become aware of the incident through their own sources and wish to participate in the investigation can also approach the investigator in charge.

   Interested parties are companies, persons, or other entities that may have an interest in the outcome in the investigation. This, of course, includes the owner, representatives of injuries or fatalities, and parties of potential civil and criminal prosecution. This may also include building component manufacturers, appliance systems providers, protection and detection system installers, equipment manufacturers, and construction material and finish manufacturers. Other parties that should not be overlooked may include service companies that serviced the equipment in the area of origin, gas and electric utilities, suppression or detection equipment representatives, representatives relating to the spread of the fire, and even some public agencies such as code enforcers or inspectors. The interested parties likely will retain their own experts, insurance companies, and lawyers that may participate in some aspect of the investigation. (page 387)

2. A work plan should be developed with the interested parties. A work plan is an outline of the tasks to be completed and includes the order and timeline for completion. The plan may involve building and site plans, manuals, and schematics to help investigators become familiar with the site. These materials can be obtained from property owners or contractors. The plan should be reviewed and modified as needed during the course of the investigation. This plan should include scene safety concerns, investigation activities, and recommendations.

   The protocol and work plan may be separate documents or combined into one document. The protocol may be a general agreement to lay out the agreement between parties of how the investigation is to proceed and the general plan and agreement.

   The work plan may be an additional document on details of each activity conducted during the investigation, the timelines for such activities, safety monitoring, and structural work. The work plan may be a more fluid document that changes as the work progresses or as issues evolve during the scene work. There may be more than one work plan that is specific to the task(s) being performed and developed as the tasks present themselves to be conducted. (page 387)

Chapter 26: Marine Fire Investigations

Matching

1. D (page 400)
2. J (page 403)
3. G (page 406)
4. H (page 403)
5. I (page 412)
6. B (page 406)
7. E (page 408)
8. A (page 406)
9. F (page 404)
10. C (page 402)

Multiple Choice

1. B (page 402)
2. C (page 401)
3. D (page 403)
4. A (page 402)
5. B (page 403)
6. C (page 405)
7. C (page 406)
8. A (page 406)
9. D (page 407)
10. B (page 408)

Fill-in

1. static discharge (page 402)
2. compartmentalized (page 402)
3. methane gas (page 402)
4. stable (page 403)
5. type of vessel (page 403)
6. cooking and heating (page 404)
7. corrosion (page 404)
8. topside (page 405)
9. hulls (page 406)
10. oxidizers (page 409)

Vocabulary

1. **Bilge:** The lowest, interior part of a vessel's hull; the area where spilled water and oil can collect. (page 403)
2. **Bulkhead:** The vertical separations in a vessel that form compartments and that correspond to walls in a building. The term does not refer to the vertical sections of a vessel's hull. (page 406)
3. **Deck:** Usually a continuous, horizontal division running the length of a vessel and extending athwart ships. This corresponds to floors in a building on land. (page 406)
4. **Hatch:** An opening in a vessel's deck that is fitted with a watertight cover. A compartment used for carrying cargo below deck in a large vessel. (page 403)
5. **Hull:** The outer skin of a vessel, including the bottom, sides, and main deck but not including the superstructure, masts, rigging, and other fittings. (page 406)
6. **Lazarette:** Stowage compartment, often in the aft end of a vessel, sometimes used as a workshop. (page 402)
7. **Races:** A component of a rolling element bearing that contains the elements and transfers the load to the bearing; generally used in pairs (inner and outer races). (page 408)
8. **Venturi:** A narrowed area within a carburetor that causes air to accelerate and creates a low-pressure area that draws fuel into the intake manifold of the engine. (page 403)

Short Answer

1. Some of the factors that can affect the interpretation of fire ignition, growth, and pattern development on a marine vessel include the following:
   - Common instances of flammable atmospheres
   - Effects on ventilation of maintaining watertight
   - Integrity at or after a fire's ignition
   - High conductivity through steel bulkheads
   - Six-way heat transfer to surrounding spaces
   - Hidden/rarely accessible spaces
   - Subsurface effects of boundary cooling
   - Movement of the shipboard platform in heavy seas

- A wide variety of hazardous materials in both the ship's stores and its cargo
- Effects of wind
- Availability of air in the fire compartment
- Use of saltwater for firefighting
- Differences in shipboard versus land-based firefighting (page 401)

2. The investigator will likely find sources of ignition similar to those found in both structural and vehicular fires:
   - Open flames
   - Electrical sources
   - Hot surfaces
   - Mechanical failures (pages 407–408)

3. As with any fire, the area of origin must be determined by the investigator and may occur in one of four major areas:
   - Cockpit/topside
   - Engine/fuel compartment
   - Accommodation compartment or cabin
   - Bilge areas (page 410)

4. Fuel tanks should be examined for failures around the edges or bottom or corrosion that may allow fugitive gases to escape. The fuel fill and vent hoses should be examined to determine whether chafing, corrosion, or other damage is present. The ground between the tank fill plate and the tank should also be tested for electrical continuity. If a fuel tank is exposed to heat, a demarcation line may be present where the fuel level inside has autocooled the exterior surface. Plastic tanks may still be intact and contain fuel. The plastic container often softens and fails at the level of the fuel within. (page 411)

5. There are literally hundreds of different types of marine vessels, each specifically designed for an intended use. Recreational or personal vessels include jet skis, wave runners, runabouts, catamarans, sailboats, and high-performance speed boats. Commercial vessels include fishing boats, container ships, ferries, recovery vessels, and oil super tanks. Military vessels include destroyers, aircraft carriers, hovercrafts, and submarines. (page 400)

Labeling

1. Basic marine vessel diagram. (page 401)

   A. Exterior.

   B. Interior.

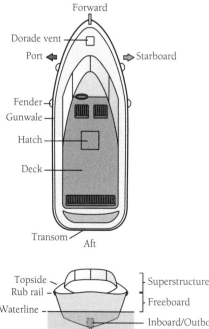

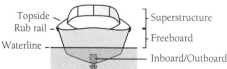

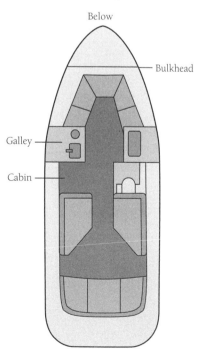

Fire Alarms

1. Either during the fire suppression operations or as a result of damage sustained because of the fire, water may enter the bilges and lower sections of the marine vessel, causing it to become very unstable. This poses the greatest hazard to the fire investigator while conducting an investigation on a marine vessel that is still floating in the water. The investigator must ensure the craft is stable before conducting the investigation because if the vessel lists or capsizes, the investigator may become trapped within it. (page 403)

2. As with any fire, the investigator should establish a basic scenario in regard to any circumstances related to the fire. The investigator should seek and interview anyone who may have information in regard to the fire, including the owner, operator at the time of the fire, bystanders, and the police and fire department personnel who responded to the incident. The investigator should seek information about the history of the vessel and any specific activities related to the fire, including activities just before, during, and after the fire occurred. The questions asked by the investigator will be very similar to questions posed during other types of fire investigation. Questions include, but are not limited to, the following topics:
   - Operational condition of the vessel
   - Accessory use and locations
   - Recent repairs or modifications
   - Weather conditions
   - Actions taken after the fire's discovery
   - Salvage operations
   - Any other questions the investigator may develop as a result of the interview itself. (page 410)